Applied Astronomy

Applied Science Review™

Astronomy

C. Gregory Seab, PhD
Associate Professor
Department of Physics
University of New Orleans
Louisiana

Springhouse Corporation
Springhouse, Pennsylvania

Staff

EXECUTIVE DIRECTOR, EDITORIAL
Stanley Loeb

SENIOR PUBLISHER, TRADE AND TEXTBOOKS
Minnie B. Rose, RN, BSN, MEd

ART DIRECTOR
John Hubbard

EDITORS
Diane Labus, David Moreau, Arielle Emmett

COPY EDITORS
Diane M. Armento, Pamela Wingrod

DESIGNERS
Stephanie Peters (associate art director), Matie Patterson (senior designer), Donald G. Knauss

ILLUSTRATORS
Philip Ashley, Robert Jackson, Stellarvisions

MANUFACTURING
Deborah Meiris (director), Anna Brindisi, Kate Davis, T.A. Landis

EDITORIAL ASSISTANTS
Caroline Lemoine, Louise Quinn, Betsy K. Snyder

Cover: *The solar system, with Jupiter in foreground.* Scott Thorn Barrows.

©1995 by Springhouse Corporation, 1111 Bethlehem Pike, P.O. Box 908, Springhouse, PA 19477-0908. All rights reserved. Reproduction in whole or part by any means whatsoever without written permission of the publisher is prohibited by law. Authorization to photocopy items for personal use, or the internal or personal use of specific clients, is granted by Springhouse Corporation for users registered with the Copyright Clearance Center (CCC) Transactional Reporting Service, provided the base fee of S00.00 per copy plus S.75 per page is paid directly to CCC, 27 Congress St., Salem, MA 01970. For those organizations that have been granted a license by CCC, a separate system of payment has been arranged. The fee code for users of the Transactional Reporting Service is 0874346061/95 S00.00 + S.75.
Printed in the United States of America.
ASR9-010794

```
Library of Congress Cataloging-in-Publication Data
  Seab, Charles Gregory
    Astronomy / C. Gregory Seab.
      p. cm. – (Applied science review)
    Includes bibliographical references and index.
  1. Astronomy.  I. Title  II. Series.
  QB43.2.S42  1995
  520—dc20                                       94-15309
  ISBN 0-87434-606-1                                  CIP
```

Contents

Advisory Board and Reviewers vi
Acknowledgments vii
Preface viii

1. **Overview of Astronomy: Observations and Measures** 1
2. **The Naked-Eye View of the Nighttime Sky** 6
3. **The Evolution of Modern Astronomy** 15
4. **Light and the Atom** 29
5. **Telescopes** 39
6. **Solar System Formation and Planetary Science** 51
7. **Earth and the Moon** 58
8. **Mercury, Venus, and Mars** 65
9. **The Giant Planets and Pluto** 73
10. **Interplanetary Bodies** 84
11. **The Sun** 91
12. **Measuring the Stars** 99
13. **Stellar Structure and Evolution** 114
14. **Stellar Remnants** 127
15. **The Milky Way** 136
16. **Galaxies** 150
17. **Cosmology** 162

Appendix: Glossary 173

Selected References 180

Index 181

Advisory Board

Leonard V. Crowley, MD
   Pathologist
   Riverside Medical Center
   Minneapolis;
   Visiting Professor
   College of St. Catherine, St. Mary's
     Campus
   Minneapolis;
   Adjunct Professor
   Lakewood Community College
   White Bear Lake, Minn.;
   Clinical Assistant Professor of Laboratory
     Medicine and Pathology
   University of Minnesota Medical School
   Minneapolis

David Garrison, PhD
   Associate Professor of Physical Therapy
   College of Allied Health
   University of Oklahoma Health Sciences
     Center
   Oklahoma City

Charlotte A. Johnston, PhD, RRA
   Chairman, Department of Health
     Information Management
   School of Allied Health Sciences
   Medical College of Georgia
   Augusta

Mary Jean Rutherford, MEd, MT(ASCP)SC
   Program Director
   Medical Technology and Medical
     Technicians—AS Programs;
   Assistant Professor in Medical Technology
   Arkansas State University
   College of Nursing and Health Professions
   State University

Jay W. Wilborn, CLS, MEd
   Director, MLT-AD Program
   Garland County Community College
   Hot Springs, Ark.

Kenneth Zwolski, RN, MS, MA, EdD
   Associate Professor
   College of New Rochelle
   School of Nursing
   New Rochelle, N.Y.

Reviewers

James MacDonald, PhD
   Associate Professor
   Department of Physics and Astronomy
   University of Delaware
   Newark

Jack W. Sulentic, PhD
   Professor of Astronomy
   Department of Physics and Astronomy
   University of Alabama
   Tuscaloosa

Acknowledgments

I gratefully acknowledge the contributions of my colleagues in astronomy, who helped me so much in the preparation of this book.

Preface

This book is one in a series designed to help students learn and study scientific concepts and essential information covered in core science subjects. Each book offers a comprehensive overview of a scientific subject as taught at the college or university level and features numerous illustrations and charts to enhance learning and studying. Each chapter includes a list of objectives, a detailed outline covering a course topic, and assorted study activities. A glossary appears at the end of each book; terms that appear in the glossary are highlighted throughout the book in boldface italic type.

Astronomy provides conceptual and factual information on the various topics covered in most introductory astronomy courses and textbooks and focuses on helping students to understand:
- principles of measuring vast distances
- the development of modern astronomy and planetary science
- the formation of the solar system
- the nature and movement of planets and interplanetary bodies (such as comets and asteroids)
- the composition and evolution of the Sun and other stars
- the nature and structure of black holes, pulsars, quasars, and other galactic phenomena
- theories of cosmology.

1

Overview of Astronomy: Observations and Measures

Objectives

After studying this chapter, the reader should be able to:
- Explain what the science of astronomy entails.
- Describe the nature of scientific theory.
- Distinguish between astronomy and astrology.
- Describe the overall structure of the universe.
- Express numerical values and carry out simple calculations using scientific notation.

I. The Science of Astronomy

A. General information
1. Astronomy is the science of all physical phenomena in the universe beyond Earth's atmosphere
2. Today astronomy is a quantitative science in which the laws of physics are applied to celestial objects and to the universe; hence the name *astrophysics* may be more appropriate and often is used
    a. Astronomy began as a purely qualitative science, meaning that no precise measurements were made
    b. While some ancient scientists (primarily the Greeks) developed quantitative methods in astronomy, the science did not become astrophysics until the time of Sir Isaac Newton in the late 17th century
        (1) Newton developed the first comprehensive mathematical basis for explaining observed phenomena
        (2) Much of Newton's epochal work was derived from his effort to explain planetary motions; thus the birth of quantitative physics has its roots in astronomy
3. Besides physics and mathematics, an astronomer may need to know chemistry, geology, atmospheric science, and certain elements of biology
    a. Astronomers study planets other than Earth, so geology and atmospheric science are required
    b. Astronomers probe molecular processes in interstellar space, in the outer atmospheres of stars, and in planetary atmospheres, so chemistry is an important discipline
    c. Astronomers are actively seeking evidence of life-forms elsewhere in the universe and therefore need some knowledge of biology

4. Approximately 5,000 professional astronomers in the United States work at universities, government research centers, or commercial industries (primarily aerospace companies)
5. Most astronomical research in the United States is supported by federal agencies, primarily the National Science Foundation (NSF) or the National Aeronautics and Space Administration (NASA)
    a. NSF funds the operations of the major U.S. observatories, including the National Optical Astronomical Observatories and the National Radio Astronomy Observatory
    b. NASA funds all space-based astronomical research, which has become an essential element in modern astronomy
6. Both on the ground and in space, astronomical research is highly dependent on technology development
    a. The most significant development for ground-based optical astronomy in recent years has been the electronic detector, particularly the Charge-Coupled Device (see Chapter 5, Telescopes, for details)
    b. The most significant benefit of space-based astronomy is the capability of observing astronomical phenomena at all wavelengths, from gamma rays to radio (see Chapter 4, Light and the Atom)

B. Astronomy and the philosophy of science
1. A scientific theory consists of a hypothesis whose predictions can be tested by experiment or observation
    a. A hypothesis may result from observation, reasoning, or even guesswork; in order to be useful scientifically, it must have observable, testable consequences
    b. In astronomy, experimentation is normally impossible because of the remoteness of the subject, and astronomers must rely almost exclusively on observation to test hypotheses
2. A scientific theory represents an attempt to find the best available explanation of a phenomenon; it is always subject to revision if observations or experiments conflict with its predictions
    a. A theory can never be proven correct but will be widely accepted if it successfully passes many tests
    b. The potential for being proven incorrect distinguishes scientific theory from faith
3. The principle known as *Occam's razor* states that the theory requiring the fewest unproven assumptions (that is, the simplest explanation) is most likely to be the correct one
    a. Thus, for example, while it is possible to construct an elaborate theory of planetary motion governed by supernatural forces, it is far simpler to accept a model in which a single, simple force — gravity — accounts for all the observed motions
    b. A more modern example is presented in cosmic background radiation (see Chapter 17, Cosmology), which can be explained simply as the remnant of an early, hot phase of the universe; earlier astronomers sought to account for cosmic background radiation in a much more complex way because they resisted the concept of an early, hot phase
4. Astrology, the belief that events in human lives on Earth are influenced by the positions and motions of heavenly bodies, originated in ancient times

a. Astrology and astronomy were indistinguishable until the 17th century, when quantitative predictions and tests of hypotheses became possible
　　　b. Astrology consistently fails to make predictions that can be verified by observation; astronomers today regard astrology as having no scientific validity

II. Overview of the Universe

A. General information
1. Earth is a tiny object in comparison with the rest of the solar system and the universe beyond
2. From Earth, with the unaided eye, we can see several types of celestial objects: the Sun, Moon, solar and lunar eclipses, five of the planets (Mercury, Venus, Mars, Jupiter, and Saturn), meteors, comets, stars, and star clusters
3. Ancient scientists and philosophers attempted to understand the universe based only on their naked-eye observations of these objects, reaching conclusions known today to be incorrect in many cases (see Chapter 2, The Naked-Eye View of the Nighttime Sky)
　　　a. For example, the ancients could discern no sense of movement of Earth, concluding that Earth was motionless at the center of the universe
　　　b. On the other hand, the ancients developed a correct explanation of eclipses of the Sun and Moon due to passage of the Moon in front of the Sun (solar eclipse) and the passage of the Moon through Earth's shadow (lunar eclipse)
4. When we look at the sky, we cannot perceive the distances of objects, but only their separation in *angular units*, defined as the degrees, minutes, or seconds of an arc; thus, we speak of the sky as if all objects were affixed to a celestial sphere, an ancient concept that remains useful today
　　　a. The ancients could measure angles to an accuracy of only a degree or so (360 degrees make up a full circle)
　　　b. In modern astronomy, scientists are often concerned with the measurement of fractions of angles as small as one second of arc, or even smaller
　　　　　(1) *Note:* There are 60 *arcseconds* in a minute of arc and 60 *arcminutes* in one degree
　　　　　(2) Consequently, there are 1,296,000 arcseconds ($360 \times 60 \times 60$) in a full circle

B. Hierarchical structure of the universe
1. Gravity governs the entire structure of the universe
2. Gravity ultimately governs the evolution of the universe; that is, the manner in which its structure changes over time
3. In our solar system, Earth and the other planets orbit the Sun, bound to it by gravitational attraction
4. The Sun is one of several hundred billion stars in the Milky Way galaxy, a vast spiral-shaped disk whose stars are bound together by gravity
　　　a. Our view of the nighttime sky is dominated by stars in the Milky Way
　　　b. The hazy band of the Milky Way is a cross-sectional, sideways view from within the disk
5. The Milky Way galaxy is one of hundreds of billions of galaxies in the universe

       a. From a vantage point randomly chosen in space, outside of any galaxy, the view would be dominated not by individual stars but by galaxies
       b. Only three galaxies — the two Magellanic Clouds and the Andromeda galaxy — are visible to the naked eye from Earth
  6. Many galaxies are grouped in clusters, in which individual galaxies are bound together by their mutual gravitational attraction
       a. The Milky Way belongs to a sparse grouping of about 30 galaxies
       b. Some clusters contain thousands of member galaxies
  7. Even the clusters of galaxies tend to be grouped into aggregations called **superclusters**, which appear to be the largest structures in the universe
       a. Superclusters tend to be filamentary or sheetlike in shape, creating an overall network of "walls" and "voids" that constitute the fundamental structure of the universe on the largest scale
       b. The origin of this structure is one of the outstanding problems in modern astronomy (see Chapter 16, Galaxies, and Chapter 17, Cosmology)
  8. There is widespread evidence that most of the mass in the universe is invisible — it is in some form other than stars or gas and dust that can be observed
       a. This **dark matter** comprises as much as 80% to 90% of all the mass in the universe
       b. The unknown nature of the dark matter is also one of the premier problems in modern astronomy (see Chapter 15, The Milky Way, and the above-named chapters)

C. The size and distance scale of the universe
  1. To discuss the vast range of sizes and distances in astronomy, astronomers use scientific notation, in which numbers are expressed in terms of powers of ten
       a. A large number, such as 47,800,000,000,000, is written as a multiplier times a positive power of ten: 4.78×10^{13}; the power of ten, or the exponent, is equal to the number of places the decimal point is moved to the left
       b. A small number, such as 0.00000000667, is written as a multiplier times a negative power of ten: 6.67×10^{-9}; the exponent is equal to the negative of the number of places the decimal point is moved to the right
       c. To multiply numbers, the product of the multipliers is taken and the exponents are added: $(4.5 \times 10^{8}) \times (3.1 \times 10^{-12}) = (13.95 \times 10^{-4})$ or (1.395×10^{-3})
       d. To divide numbers, the quotient of the multipliers is taken and the exponents are subtracted: $(7.94 \times 10^{23}) \div (1.12 \times 10^{7}) = (7.09 \times 10^{16})$
  2. Astronomers must understand the nature of *atoms* and *molecules*, whose sizes are typically 10^{-10} to 10^{-9} m, as well as planets (10^{6} to 10^{7} m), stars (10^{9} to 10^{11} m), interstellar distances (10^{16} to 10^{18} m), galaxies (10^{20} to 10^{21} m), intergalactic distances (10^{22} to 10^{23} m), and the universe itself (10^{25} m)
  3. The universe consists mostly of empty space: to visualize this, imagine that Earth is scaled down to the size of a basketball (0.3 m in diameter); the Sun, which is 109 times larger (or about 33 m in diameter on this scale), would then be over 3.5 km away, and the nearest star would be over 1 million km away
  4. In the "official" unit system used by astronomers and physicists, called the *Systeme Internationale*, distances are expressed in meters, masses (discussed in Chapter 3, The Evolution of Modern Astronomy) are measured in kilograms, and time is measured in seconds

5. Astronomers commonly use special units that make values easier to express or comparisons easier to understand
   a. The **astronomical unit (AU)**, the average distance between the Sun and Earth, is convenient for expressing distances within the solar system; 1 AU = 1.5×10^{11} m or 1.5×10^{8} km
   b. The properties of stars are often stated in terms based on measurements of the Sun; for example, an astronomer might describe a certain star as having a mass of 5 solar masses and a radius of 2 solar radii, rather than expressing its mass in kilograms and its radius in meters
   c. Large distances are sometimes expressed in terms of the *light-year* — the distance light travels (in a vacuum) in one year at its speed of 3×10^{8} m/sec; one light-year equals about 9.5×10^{15} m (another distance unit, the **parsec**, equals 3.26 light-years)

Study Activities

1. In medieval times, astronomy was supported almost exclusively by colleges and by monarchs (that is, many kings or queens appointed and supported royal astronomers). Comment on the parallels between that situation and the present-day support of astronomy.
2. Compare a belief system of your choice (a modern religion; astrology; a fringe science, such as iridology or therapeutic touch; or a mainstream scientific theory) with the standards set forth in this chapter for a scientific theory. In your discussion, comment on the following: Does your featured belief system yield predictions that can be tested? Can it be proven wrong? If predictions and tests are possible, have they been performed? If so, what was the result? Does your chosen belief system qualify as a scientific theory?
3. Explain why it is convenient for astronomers to speak of the sky as though all the objects visible were fixed to a sphere centered on Earth.
4. The visible planets all move across the sky along a common path. By the same token, the Milky Way stretches across the sky, a band of stars. Both phenomena are concentrated in well-defined strips across the sky. What does this tell us about the structure of the solar system, and of the Milky Way?
5. Describe the overall structure of the universe, starting on the largest scale and proceeding toward smaller structures.
6. Explain how gravity influences the fate and structure of the universe.
7. Write the following numbers using scientific (powers-of-ten) notation:
   (a) 45,670,000
   (b) 0.00000000000000000000056.
8. Calculate the relative distance between the Sun and Earth if Earth were shrunk to 1 mm in diameter, and the Sun's size and distance were scaled accordingly. What is the Sun's diameter on this scale?

2

The Naked-Eye View of the Nighttime Sky

Objectives

After studying this chapter, the reader should be able to:
- Explain the daily motions of objects in the nighttime sky.
- Describe the coordinate system most commonly used in astronomy for locating objects in the sky.
- Explain seasons on Earth.
- Discuss the appearance and motions of the Moon.
- Explain eclipses of the Sun and Moon.
- Describe the motions of the visible planets.
- Summarize the early historical development of astronomy.

I. Apparent Motions Due to Motions of Earth

A. General information
1. Earth spins (rotates) as it revolves around the Sun in its orbit
2. The rotation of Earth causes daily, or diurnal, motions as all objects in the sky rise and set
3. The revolution of Earth about the Sun causes annual phenomena, such as the seasons and the progression of the Sun through the constellations

B. Daily (diurnal) motions
1. As seen from the surface of Earth, all objects in the sky rise in the east and set in the west every day; this is an apparent motion because we are observing the sky from a spinning platform, Earth
   a. The points in the sky directly above the north and south poles are the *celestial poles;* the sky appears to rotate about these points
   b. Objects near the poles never set, because they stay above the horizon throughout the day and night
      (1) These *circumpolar* objects appear to circle the pole as Earth spins
      (2) The size of the circumpolar region seen from any given location on Earth depends on how close to the pole that location is; at the pole, for example, the entire sky appears to circle without setting, whereas from the equator only stars right at the pole position can be seen at any time of the night
2. The length of day depends on whether it is measured by an observer on the spinning Earth or by an observer located in space at a fixed position relative to the background stars (see *Solar Day vs. Sidereal Day*)

Solar Day vs. Sidereal Day

The arrow pointing toward the Sun indicates the overhead direction from a fixed point on Earth. A sidereal day is a time interval defined from the direction of the arrow at noon one day (left) to the moment the arrow is pointing in the same direction the next day, as seen by a distant observer. Because Earth has moved, however, it will be about 4 minutes later when the arrow points directly at the Sun again (right); hence the sidereal day is nearly 4 minutes shorter than the solar day.

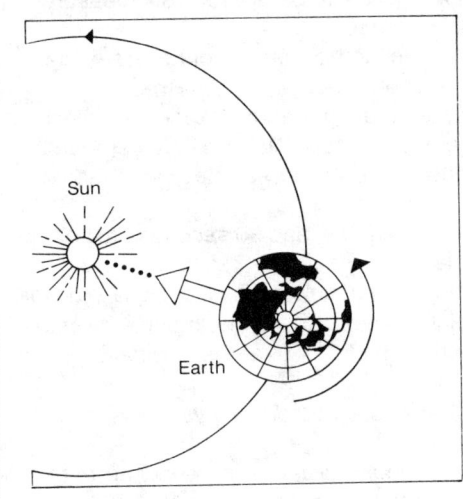
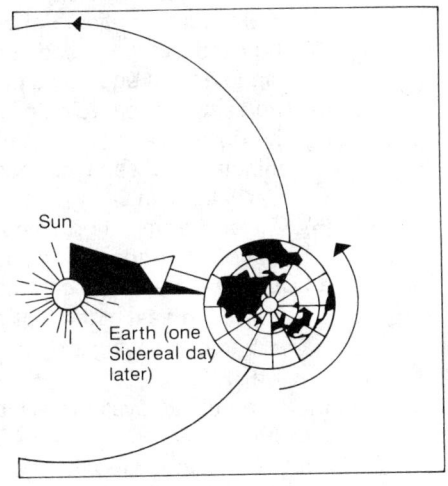

    a. The **solar day** is the length of day as observed by a person on the spinning Earth; for example, the time from noon one day to noon the next
    b. The **sidereal day** is the length of time it takes for Earth to complete one full rotation, as seen by an observer in space
    c. Because of Earth's orbital motion around the Sun, the solar day is a little longer (about 4 minutes) than the sidereal day
        (1) As Earth moves along its orbit during a day, the direction to the Sun gradually changes, and Earth must complete more than a full rotation in order to return to the same orientation in relation to the Sun
        (2) Timekeeping is based on the solar day rather than the sidereal day because the solar day is apparent to people on Earth's surface
3. More specifically, timekeeping is based on the average length of the solar day, called the **mean solar day**, which is equal to exactly 24 hours (the length of the solar day actually varies a little during the year because Earth's orbital speed is not constant; this is why the average length is used)
4. Modern clocks, such as atomic clocks, can measure relative time very accurately, but because Earth's rotation rate varies from time to time, occasional adjustments must be made to "official time," often in the form of *leap seconds*

C. Astronomical coordinate systems
    1. The most commonly used astronomical coordinate system for measuring locations in the sky is the **equatorial coordinate system**
        a. This system is essentially equivalent to the latitude and longitude system used for measuring positions on Earth's surface

b. The north-south coordinate (the analog of latitude) is the angular distance of an object north or south of the *celestial equator,* the projection of Earth's equator onto the sky
   (1) This coordinate is called the **declination** of an object
   (2) Declinations are expressed in units of angle measurement (degrees, minutes, and seconds)
c. The east-west coordinate is measured from a fixed direction in the sky
   (1) Because Earth rotates, it is not practical to base east-west measurements on a location on Earth's surface
   (2) The fixed direction used in space is the intersection of Earth's orbital plane and its equatorial plane, called the *first point in Aries*
   (3) The east-west coordinate is called the **right ascension** of an object
   (4) The right ascension of an object is measured in units of time (hours, minutes, and seconds) because the entire sky rotates overhead as Earth turns in 24 hours
   (5) A given star might be 14 hours, 6 minutes, and 29 seconds east of the first point in Aries, for example
   (6) Using the equatorial coordinate system, astronomers designate a star 90° to the east of the first point in Aries and 40° north of the celestial equator as having a right ascension of $6^h0^m0^s$ and a declination of +40°
2. An alternate coordinate system, called the *ecliptic coordinate system,* is sometimes useful
   a. Ecliptic coordinates are based on the angular distance in the sky north or south of the plane of Earth's orbit (called the **ecliptic**) and east of the Sun's position
   b. This coordinate system is especially useful in studies of the solar system, because it relates a celestial body's position to its location within the solar system
3. Another useful coordinate system is the *galactic coordinate system*
   a. In this system a body's position is expressed in angular units relative to the plane of the galactic disk and the direction of the galactic center
   b. This system is especially useful in studies related to the structure of the galaxy, because it expresses positions relative to the galaxy

D. Annual motions and the seasons
1. As Earth orbits the Sun, it appears to us that the Sun moves through the background of stars, taking one year to make a complete circuit
2. The precise line followed by the Sun, which represents the plane of Earth's orbit, is called the **ecliptic**
3. The broad band of constellations through which the Sun appears to move is called the **zodiac**
   a. Because the planets follow orbits in planes nearly aligned with the ecliptic, they, too, appear to move through the zodiac
   b. Ancient observers studied these apparent motions through the zodiac, giving rise to theories of astrology
4. *Seasons* are caused by the tipping of Earth's rotation axis 23.5° away from the perpendicular to its orbital plane; this orientation remains more or less constant as Earth orbits the Sun

      a. As Earth orbits the Sun, the northern and southern hemispheres are alternately tipped toward the Sun and then, 6 months later, away from it
      b. When one hemisphere is tipped toward the Sun, that hemisphere has summer
          (1) Summer weather is warmer because the Sun's rays strike at more nearly perpendicular angles than in the winter; thus the Sun's intensity at Earth's surface is greater during summer
          (2) Another reason for summer's warm weather is that the days are longer, because sunlight extends over more than half of the hemisphere tipped toward the Sun
      c. The hemisphere tipped away from the Sun has winter
5. Latitude zones on Earth's surface are defined according to the apparent position of the Sun at different times of the year
      a. Because of the tilt of Earth's axis, it appears to us that the Sun moves north and south during the year, reaching maximum declinations of 23.5° N and 23.5° S
          (1) Consequently, the Sun can be directly overhead only at latitudes between 23.5° N and 23.5° S
          (2) This range of latitudes on Earth defines the tropics
      b. The arctic circle and the antarctic circle are the regions around the poles illuminated throughout the 24-hour day when the Sun is at its maximum northerly or southerly positions
          (1) The arctic zones extend 23.5° from the poles; they lie at latitudes above 66.5° N and below 66.5° S
          (2) These zones lie in darkness throughout the 24-hour day in winter, when the Sun is at its maximum declination in the opposite hemisphere
6. The times when the Sun reaches its maximum distances north or south of the equator are called **solstices**
      a. These occur on or about June 21 and December 21 each year
      b. Solstices are celebrated throughout the world because they signify the dates when days reach their maximum length and begin to shorten, or reach their minimum length and begin to lengthen
7. The times when the Sun "crosses" the equator are called **equinoxes**
      a. These occur on or about March 21 and September 22 each year
      b. At these times the lengths of day and night are equal because the Sun stays over the equator as Earth spins
      c. At the *vernal equinox*, occurring in the northern spring, the Sun crosses the equator from south to north
          (1) At this time the Sun is at the position where Earth's orbital and equatorial planes intersect
          (2) Thus the Sun is at the position of the first point in Aries (the zero point for right ascension, as described earlier)
      d. At the *autumnal equinox*, in the northern fall, the Sun crosses the equator from north to south
8. Earth undergoes a slow wobbling motion, called **precession**, much like a toy top
      a. It takes 26,000 years for Earth to undergo one cycle of precession
      b. Precession causes the coordinates of objects in the equatorial coordinate system to change gradually; thus astronomers must correct for these shifts when making observations

c. Over the centuries, precession also has caused a gradual shifting of the Sun's position in the zodiac
   (1) The Sun today is consequently not in the same constellation at a given time of year as it was hundreds of years ago or will be hundreds of years from now
   (2) One consequence is that the Sun today is never in the constellation associated with a given date by astrology; for example, the "Sun signs" of the zodiac refer to a time some 2,000 years ago and are no longer correct, and the first point of Aries has shifted into Pisces
   (3) Another consequence is that the pole star Polaris, now located almost exactly at the north celestial pole, is only the pole star temporarily; normally there is no pole star

II. Lunar Motion and Eclipses of the Sun and Moon

A. General information
1. The Moon orbits Earth while Earth orbits the Sun, causing the Moon to undergo changes in appearance as seen from Earth
2. The Moon emits no light of its own; it is seen only by reflected sunlight
3. Because Earth moves along its orbit at the same time the Moon orbits Earth, the length of the Moon's orbital period depends on whether the observer is on Earth or at a fixed location in space (see *Sidereal and Synodic Periods of the Moon*)
   a. From our position on the moving Earth, we observe the **synodic period** or **lunar month** of approximately 29.5 days
   b. The length of the Moon's orbital period as seen by a distant observer is the **sidereal period**, which has a length of about 27.3 days
   c. As seen by an outside observer, the Moon completes more than one full circle in order to go from a full-moon phase to the next; hence the lunar (or synodic) month is about 2 days longer than the Moon's sidereal period
4. Ancient scientists and philosophers understood lunar motion well because the Moon is the one body that truly orbits Earth

B. Phases and configurations of the Moon
1. During one lunar month (that is, the synodic period of the Moon), the Moon moves through a sequence of *configurations*, or specific positions relative to the Sun-Earth direction (see *Lunar Phases and Configurations*, page 12)
   a. When the Moon is opposite the Sun's direction, it is at *opposition*
   b. When the Moon is aligned with the Sun, it is at *conjunction* (and cannot be observed because the Sun's brilliance conceals it)
   c. When the Moon is at a 90° angle with respect to the Earth-Sun line, it is at *quadrature*
2. As the Moon moves through its sequence of configurations, its appearance, or *lunar phase*, changes; this refers to the shape of the sunlit portion as seen from Earth
   a. When at opposition, we see one entire hemisphere of the Moon illuminated; this is the *full-moon* phase
   b. When at conjunction, the Moon presents its unlit side to Earth and is not visible; this is the *new-moon* phase

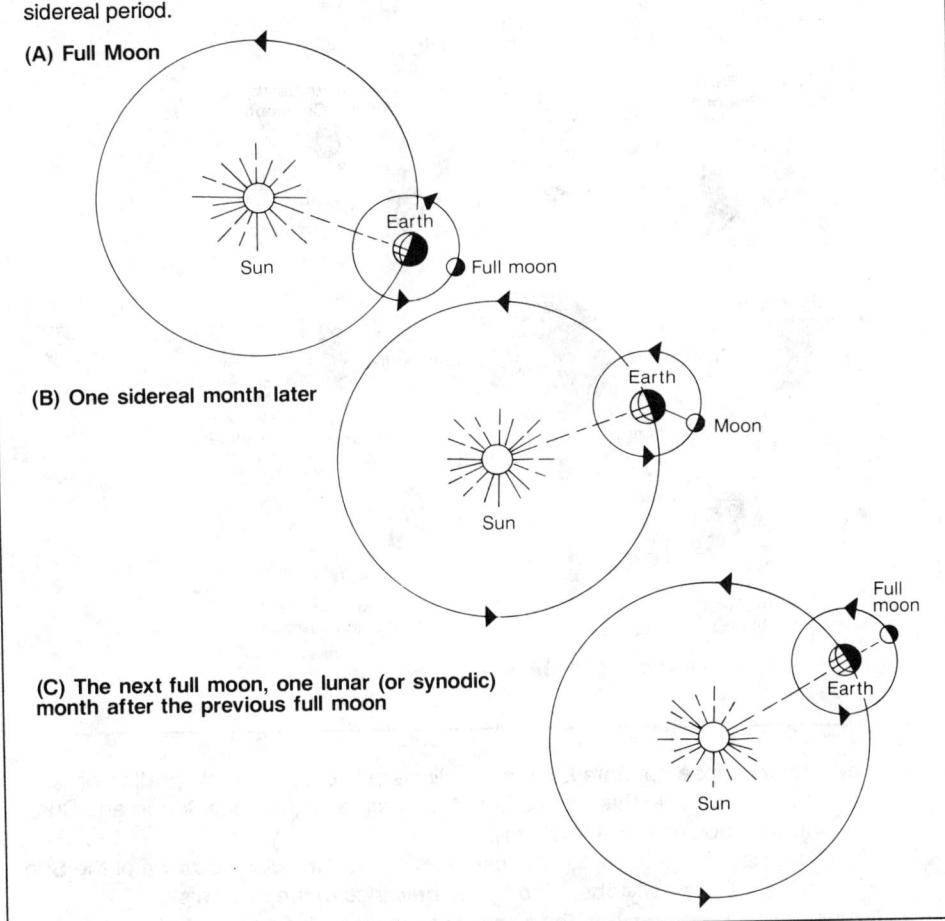

Sidereal and Synodic Periods of the Moon

As seen by an observer in space, the Moon goes through more than one full circle when traveling from one full moon to the next. This is because Earth is moving in its orbit while the Moon orbits it. Hence, the lunar (or synodic) month is about 2 days longer than the Moon's sidereal period.

(A) Full Moon

(B) One sidereal month later

(C) The next full moon, one lunar (or synodic) month after the previous full moon

    c. When the Moon is at quadrature, we see one-half of its sunlit side, or a half circle; the two phases where this occurs are called *first-quarter* and *third-quarter* moon

C. Lunar and solar eclipses
1. Eclipses of the Moon or the Sun occur as our view of one is obscured by the other or by Earth (see *Solar and Lunar Eclipses,* page 13)
2. A *solar eclipse* occurs when the Moon moves directly between Earth and the Sun, blocking some or all of the Sun's disk from view
    a. The eclipse is total if the full disk of the Sun is blocked
    b. It is entirely coincidental that the Sun and the Moon have disks having the same angular size (about 30 arcminutes) as seen from Earth; the Sun is actually much larger than the Moon but also is much farther away

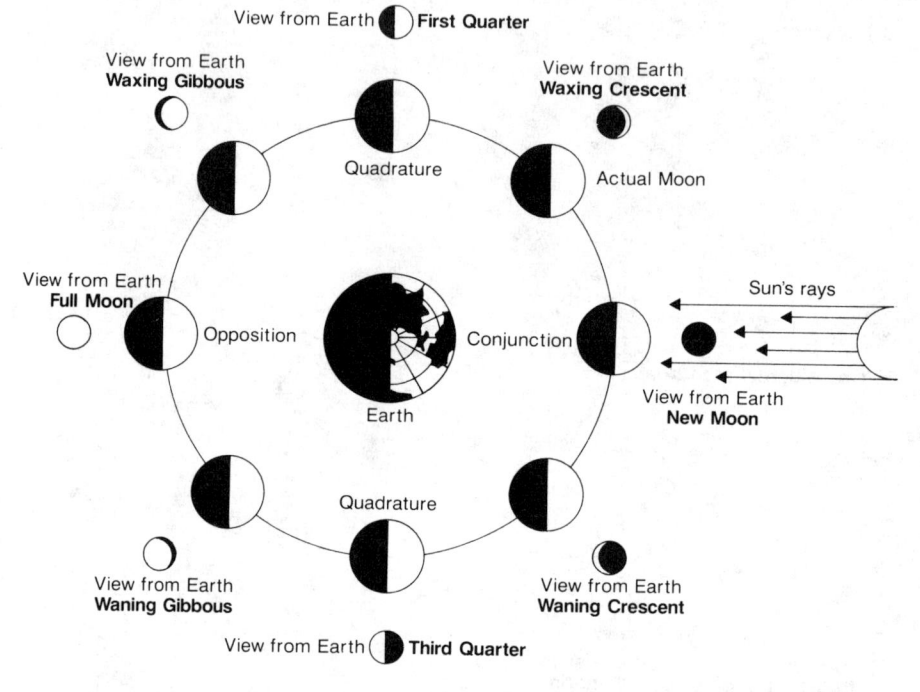

Lunar Phases and Configurations

Varying portions of the Moon's sunlit side can be seen during orbit. The diagrams outside the circle (representing the Moon's orbit) show the lunar phases as they appear to observers from the northern hemisphere.

    c. A solar eclipse appears as a total eclipse only over a narrow portion of Earth's surface; this results from the precise alignment of Moon and Sun that is required for a total eclipse

    d. During a solar eclipse, Earth observers see a dim, wispy **corona** of the Sun, which is normally obscured by the brilliance of the solar disk

3. A *lunar eclipse* occurs when the Moon passes through Earth's shadow

    a. A lunar eclipse is total if the entire Moon is immersed in the dark inner portion of Earth's shadow

    b. Earth's shadow at the Moon's distance from Earth is more than twice as large as the Moon, so lunar eclipses do not require a very precise alignment of Sun, Earth, and Moon

4. Eclipses do not occur every lunar month because the Moon's orbital plane is tipped (by about 5°) with respect to Earth's orbit

    a. Thus, on most orbits, the Moon passes above or below Earth's shadow, and no eclipse occurs

    b. The plane of the Moon's orbit precesses (wobbles) with a period of about 18 years

        (1) Consequently, every 18 years the pattern of eclipses repeats itself, as the line of intersection between the Moon and Earth's orbital planes returns to the same alignment

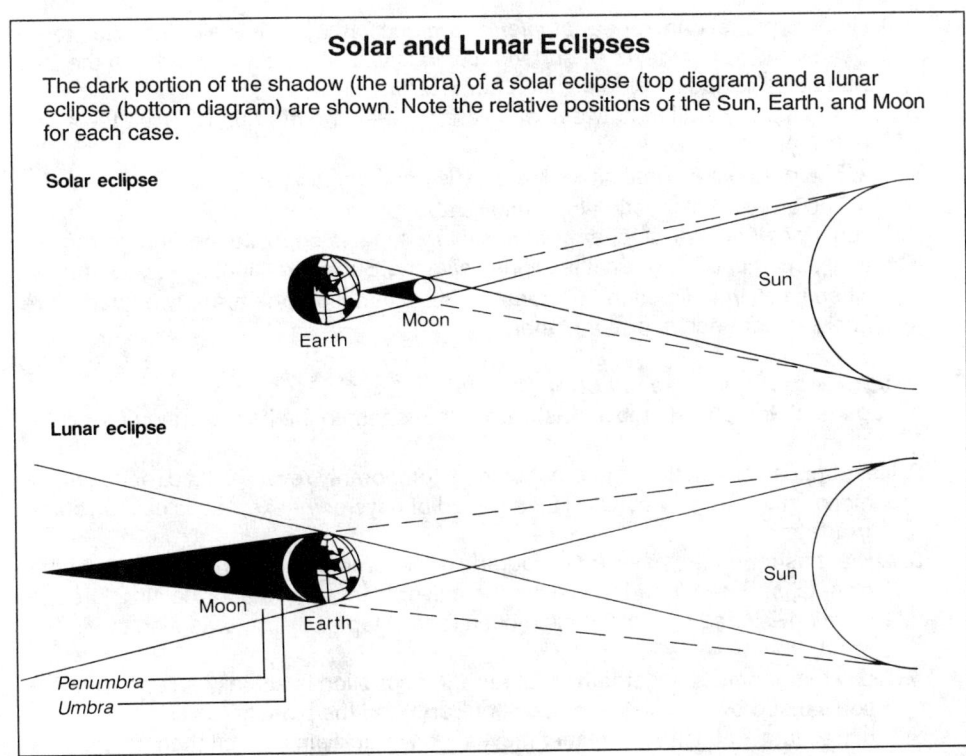

Solar and Lunar Eclipses

The dark portion of the shadow (the umbra) of a solar eclipse (top diagram) and a lunar eclipse (bottom diagram) are shown. Note the relative positions of the Sun, Earth, and Moon for each case.

        (2) This 18-year pattern of eclipses, called the *saros*, was recognized in ancient times
    5. Solar eclipses are much less frequent than lunar eclipses, because a solar eclipse requires a more precise alignment of Earth, Sun, and Moon than a lunar eclipse

III. Planetary Motions

A. General information
    1. The nine planets of the solar system all orbit the Sun in the same direction, with orbital planes that are nearly aligned; the entire system has a disklike structure
    2. The orbital period observed for a planet depends on the location of the observer
        a. From Earth, which moves at the same time a planet is being observed, we measure the *synodic period* of a planet
        b. From a fixed location in space, we measure the *sidereal period*
    3. Each planet in succession outward from the Sun moves more slowly in its orbit
    4. Planets closer to the Sun than Earth are called *inferior planets;* those farther from the Sun than Earth are called *superior planets*

B. Configurations of the planets
    1. Each planet moves through a sequence of *configurations*, depending on its position relative to the Sun-Earth line

2. An inferior planet can appear at *inferior conjunction*, when it is aligned with the Sun on the near side, or at *superior conjunction*, when it is aligned with the Sun but is on the far side as seen from Earth
3. The angular separation between an inferior planet and the Sun is called the *elongation* of the planet
   a. The greatest elongation for Mercury is approximately 28°
   b. The greatest elongation for Venus is about 47°
4. A superior planet can appear at *opposition* when it is opposite the Sun's direction; at *conjunction* when it is aligned with the Sun's direction; or at *quadrature* when it lies in a direction 90° from the Sun's direction (there are two quadrature positions for each superior planet)

C. Retrograde motion of the superior planets
1. In general, the planets move eastward with respect to the background of fixed stars
2. Near opposition, each of the outer planets temporarily reverses its direction of motion and moves westward for a period of days or weeks; this is called **retrograde motion**
3. Ancient astronomers believed retrograde motion represented a real motion of the planets, and developed models of the universe in which elaborate structures were envisioned to explain this motion (see Chapter 3, The Evolution of Modern Astronomy)
4. Today astronomers understand that retrograde motion is actually an apparent motion caused by the relative motions of Earth and the planets
   a. Because each inferior planet moves more quickly in its orbit than the planets farther from the Sun, Earth periodically overtakes and passes each of the superior planets
   b. As Earth passes a superior planet, our line of sight temporarily sweeps backward with respect to the distant stars, causing the planet to appear as though it were temporarily moving backward
5. The inferior planets undergo retrograde motion when they are on the far side of the Sun, apparently going in the opposite direction from that of Earth

---

Study Activities
1. Make a sketch showing the motion and spin of Earth as it orbits the Sun. Explain why the solar day is a little longer than the sidereal day.
2. Explain why the mean solar day is the basis for timekeeping, instead of the sidereal day or the true solar day (that is, the actual length of each day).
3. Explain the distinction between the ecliptic and the zodiac.
4. Describe what causes the seasons and why summer is so much warmer than winter at temperate latitudes.
5. Explain why the position of the Sun relative to the zodiac differs from the astrological sign for a given date.
6. Make a sketch to explain the difference between the synodic period of the Moon (that is, the lunar month) and the Moon's sidereal period.
7. Describe the retrograde motion of an outer planet.

3

The Evolution of Modern Astronomy

Objectives

After studying this chapter, the reader should be able to:
- Summarize the early development of scientific thought in ancient Greece.
- Trace the development of astronomy from its beginnings in the Middle East through the culmination of the Greek civilization.
- Describe the developments of the Renaissance, when astronomy evolved from the Earth-centered view of the Greeks to the modern Sun-centered picture, through the efforts of Copernicus, Brahe, Kepler, Galileo, Newton, and others.
- Describe and explain the laws of motion and gravitation developed by Newton.
- Discuss the principles of energy and orbits.
- Explain tides.

I. Early Development of the Oldest Science

A. General information
1. Astronomical knowledge developed independently in several parts of the world, but modern astronomy traces its roots to Mediterranean origins, in what is now Iraq, Greece, and Egypt
2. The ancient observers were aware of many phenomena that could be observed with the naked eye
    a. The length of the year was measured accurately as long ago as 3,000 B.C.
    b. The correct explanation of eclipses was known as early as 500 B.C.
    c. The cycles of lunar and planetary motions were well known from early times
    d. The Greeks were aware of the 18-year pattern of eclipses more than 2,000 years ago
    e. Even precession, the subtle shifting of star positions over the years, was discovered by Hipparchus around 150 B.C.
3. **Cosmology** is the study of the structure and evolution of the universe
    a. Some ancient cultures developed cosmologies while others dealt only with what they could observe, making few attempts to explain the phenomena
    b. The early Greeks developed the first cosmologies that attempted to tie known laws of nature to an understanding of the universe; thus modern science is descended most directly from the Greeks

B. The Babylonians
1. The Babylonian civilization flourished for many centuries in what is now Iraq, beginning around 2,000 B.C.

2. Babylonian astronomers made many careful measurements of the skies and knew the length of the year to an accuracy within a few minutes of the modern value
   a. This was done by noting that the length of the shadow of a fixed stake changes during the year, as the Sun's altitude above the horizon varies
   b. To measure the length of the year requires simply counting the days required for the length of the shadow to progress through a full cycle of changing length
3. The Babylonians developed a 360-day calendar containing twelve 30-day months (occasionally an extra month had to be added to compensate for the 5.25 days that were omitted each year)
4. The full circle of the sky was divided into 360 parts, corresponding to the Sun's position over the 360 days of the year
   a. Thus the Babylonians originated the modern system of *angular measure,* with 360 degrees in a full circle
   b. The Babylonians thought the number 60 was especially significant, because it divided evenly into 360 and was evenly divisible by 12, the number of months in the year
      (1) Therefore, each degree has 60 parts *(arcminutes)*, and each arcminute has 60 parts *(arcseconds)*
      (2) This also explains why the Babylonian timekeeping system has 60 minutes in an hour, and 60 seconds in a minute of time
5. Later civilizations in the region, such as the Assyrians and the Chaldeans, maintained the Babylonian records and teachings, and added their own knowledge
   a. The Chaldeans, for example, developed tables from which the motions of the Sun and the Moon, as well as eclipses, could be predicted
   b. Gradually the center of influence and knowledge shifted to the west, toward the shores of the Mediterranean

C. The Greeks
1. The earliest Greek civilization arose on the island of Crete, from about 5,000 B.C. to 2,000 B.C.; the mythology of the constellations is attributed to the early Minoan culture on Crete
2. The beginnings of formal scientific thought are traced to the Greek philosopher Thales (circa 624-547 B.C.), who taught that rational inquiry can lead to understanding
3. Later, Pythagoras (circa 570-500 B.C.) developed the school of thought that reality can be represented by numbers
   a. Pythagoras laid the foundations for geometry and trigonometry
   b. He also is credited with being the first to state that Earth is spherical, and that all motions of heavenly bodies are perfectly circular (now known to be untrue)
   c. Pythagoras also introduced the concept that the planets created celestial music as they moved, with harmonies dictated by their relative distances
4. One of the most influential of the ancient Greeks was Plato (428-347 B.C.), who taught that what we see of the universe is an imperfect representation of the underlying perfection
   a. One corollary to Plato's theory is that observations are not necessary because one can deduce the nature of the universe by reason and deduction

Early Development of the Oldest Science 17

      b. Plato's deductive approach, in which observation and experiment were largely ignored, had a great deal of influence on later philosophers
5. Plato's most important follower was Aristotle (circa 384-322 B.C.)
      a. Aristotle expanded on Plato's teachings by invoking a system of physical laws and using them to deduce properties of the universe
         (1) This was important because he gave reasons for his beliefs, instead of simply stating them without support
         (2) As in Plato's case, Aristotle developed his basic principles by logical deduction, rather than basing them on observation and experiment
      b. Using his laws, Aristotle could demonstrate that Earth is spherical
      c. Aristotle also taught that all heavenly bodies were perfect, unchanging spheres, and that all moved in perfect circles
      d. Aristotle based his cosmology on the assumption that Earth lies at the center of the universe; this concept, along with the belief that all motions in the heavens are perfectly circular, dominated cosmological thinking for some 2,000 years
6. The center of Greek culture shifted across the Mediterranean to Alexandria during the fourth century B.C.
7. The first great figure after that time was Aristarchus (circa 310-230 B.C.), who deduced that Earth must orbit the Sun
      a. Aristarchus used geometric arguments to show that the Sun is much larger than Earth and must therefore be the central body in the universe
      b. This notion did not win great acceptance at the time
         (1) The Sun-centered theory was not accepted because people did not perceive any shortcomings with the standard Earth-centered view taught by Aristotle and his followers
         (2) Another reason is that no **stellar parallax** was observed; this is the apparent shifting of position of nearby stars due to Earth's orbital motion (described in Chapter 12, Measuring the Stars)
8. At about the same time as the work of Aristarchus, Eratosthenes (circa 273 B.C.) used geometric arguments to measure the size of Earth to within 2 percent of the modern value
      a. Eratosthenes measured the direction toward the Sun on the same day, from two different points on Earth about 500 miles apart
      b. He reasoned that the separation of these two points on Earth's surface must represent the same fraction of Earth's circumference as the ratio of the angular difference to a full circle
9. The concept of the epicycle was introduced by Apollonius (circa 265-190 B.C.), who developed it as a means of explaining the temporary retrograde (backward) motions of the planets (see Chapter 2, The Naked-Eye View of the Nighttime Sky)
      a. An *epicycle* is a small circle on which a planet moved, which itself moved around Earth on a larger circle
      b. The epicycle is another concept of the ancient Greeks that persisted for a very long time
10. The epicyclic theory was developed in some detail by Hipparchus (circa 150 B.C.)
      a. In his epicycle model, Hipparchus found it necessary to violate some of the detailed teachings of Aristotle
         (1) For example, in order to account for observed variations in the apparent speed of the planets, it was necessary to shift Earth slightly off-

center within the large circles on which the epicycles of the planets revolved
            (2) In doing so, Hipparchus departed from the Aristotelian philosophy that observations should be ignored if they are not perfectly consistent with theory
        b. Hipparchus made many other important contributions, including the first application of trigonometry to the skies; the development of a star catalog; refinement of astronomical instruments; and the invention of the **stellar magnitude system** for measuring stellar brightness (see Chapter 12, Measuring the Stars)
        c. He discovered **precession** (accomplished by comparing his measurements of star positions with those made over a century earlier by other scientists)
    11. Ptolemy (circa 150 A.D.) was the last great scientist of the Greek era
        a. He collected all the world's knowledge of astronomy in a 13-volume work called the *Almagest*
        b. Ptolemy also made his own refinements in instrumentation and in the epicyclic cosmology

D. Other cultures
   1. Many of the developments of the Greeks were paralleled by advances made elsewhere; however, only the Greeks developed a sophisticated cosmology linking the structure of the universe to known physical laws.
        a. The Chinese, whose astronomical traditions date back at least to 2,000 B.C., knew the length of the year and the lunar month to a high degree of accuracy, and developed a star catalog similar to that of Hipparchus
        b. Starting around 1,500 B.C., Hindu astronomers in India developed a very sophisticated calendar recognizing several long-term cycles in lunar and solar motions that were not noticed by the Greeks
        c. Astronomers in Central and South America developed complex calendars and constructed their entire cultures around astronomically significant events and alignments
        d. In North America, Native American cultures had a rich oral tradition recognizing important astronomical cycles, and many monuments were constructed incorporating astronomical alignments
    2. It is not known how much communication and interaction occurred between these cultures and the Greek civilization

II. The Renaissance

A. General information
   1. Following the fall of the Greek empire, astronomical traditions and lore were preserved by the Arabs who occupied northern Africa and southern Europe
   2. Few additional astronomical developments occurred for over a thousand years following the time of Ptolemy
   3. In the 15th century, an intellectual, artistic, and cultural re-awakening known as the Renaissance began in Europe
        a. The dominant theme of this era was religious rebellion, as Protestant reforms swept Europe

  b. Along with this major cultural revolution came advances in philosophy, mathematics, and science

B. Nicolaus Copernicus (1473 - 1543)
1. Copernicus was born and educated in Poland; in his university studies of astronomy, he was taught the standard doctrine, which dated all the way back to Aristotle, Hipparchus, and Ptolemy
2. Copernicus came to believe that the Sun, not Earth, lay at the center of the universe; his principal reasons were aesthetic and philosophical, not because his model was more accurate than those of his predecessors
  a. He was taken by the aesthetic appeal of the concentric pattern of planetary orbits in the Sun-centered (*heliocentric*) model, in which the relative planetary distances from the Sun could be deduced
  b. His mathematical model was no more accurate than that of Ptolemy
  c. Copernicus was forced to introduce epicycles to account for some of the irregularities of planetary speeds and distances
  d. The Copernican model did enjoy some successes over the Earth-centered cosmology
    (1) Copernicus correctly explained the cause of the seasons as being due to a tilt in Earth's axis
    (2) He correctly explained the reason for retrograde motion of the planets
    (3) He was able to determine the relative distances of the planets from the Sun in his model, whereas in the Earth-centered cosmology, planetary distances were free parameters
3. Copernicus was reluctant to publish his ideas, and only did so just before his death; he did not live to see the impact of his theory

C. Tycho Brahe (1546 - 1601)
1. Tycho Brahe lived in Denmark and was trained as a mathematician and astronomer
2. He achieved fame through his observations and analyses of a supernova and a comet
3. In 1575, Brahe was appointed royal astronomer by King Frederick II of Denmark and given an island on which to build an observatory
4. Brahe developed large instruments with accurately calibrated scales, thus improving the accuracy of positional measurements
5. Brahe recorded positions of planets at every opportunity, rather than doing so only at significant times, such as oppositions and conjunctions, which had been standard practice
  a. He made multiple measurements of a planet's position each time to reduce the chance for random error
  b. Averaging together the multiple measurements yielded greatly improved accuracy; this technique commonly is used by modern scientists
6. Brahe was unable to accept the Sun-centered cosmology, suggesting instead that the Sun orbited Earth, but that all the other planets orbited the Sun; these theories were not well accepted

D. Johannes Kepler (1571 - 1630)
1. Kepler lived in what is now Germany and was exposed to astronomy while studying theology at a university

2. He had a deeply held belief that the cosmos was arranged according to mathematical principles; he believed in heliocentric cosmology once he learned the Copernican theory
3. Kepler began working for Tycho Brahe shortly before Brahe's death
   a. Brahe assigned Kepler the task of developing an accurate understanding of planetary motions
   b. When Brahe died, Kepler was given access to Brahe's complete archive of observational data and appointed by the Danish King to continue his task
4. Kepler discovered several important relationships describing the orbits of the planets
   a. He initially worked with the data about Mars, the best observed of all the planets; the orbit of Mars departs from a perfect circle more significantly than do the orbits of the other planets for which he had extensive data
   b. Eventually he showed that the other planetary orbits behaved similarly to that of Mars
5. Kepler's *first law of planetary motion*, discovered after much experimentation, states that the orbit of each planet is an ellipse with the Sun at one focus (see *Kepler's First and Second Laws*)
   a. An ellipse is an oval figure such that the sum of the distances from any point on it to two fixed points (called *foci*) is constant
   b. The Sun is located at one focus of the ellipse representing a planetary orbit; the other focus is empty
6. Kepler's *second law of planetary motion* states that a line connecting a planet to the Sun sweeps out equal areas in space in equal intervals of time (see *Kepler's First and Second Laws*)
   a. This means that a planet moves fastest when closest to the Sun and slowest when farthest away
   b. A planet's closest approach to the Sun is called the *perihelion;* its farthest approach, the *aphelion*
   c. In modern times, it is recognized that *angular momentum* in a two-body orbital system is constant
7. Kepler's *third law of planetary motion* relates the size of a planet's orbit to its sidereal period, stating that the square of the sidereal period is proportional to the cube of the semimajor axis
   a. The *semimajor axis* is a measure of the size of the orbit; this is equal to one-half of the long axis of the ellipse
      (1) Thus, the semimajor axis represents an average distance of the planet from the Sun
      (2) For a circular orbit, the semimajor axis is the same as the radius of the circle
   b. If the semimajor axis (a) is measured in astronomical units and the orbital period (P) in years, then Kepler found that $P^2 = a^3$
8. Using his laws, Kepler was able to improve, by a factor of more than 100, the accuracy of tables predicting planetary motion
   a. This represented a resounding confirmation of the Sun-centered cosmology, since no Earth-centered theory could approach the same level of accuracy
   b. By the time of Kepler's death, his work, along with that of Galileo, had effectively invalidated the Aristotelian model of the universe

Kepler's First and Second Laws

Kepler's first law states that planetary orbits are ellipses with the Sun off-center at one focus, as shown in the illustration on the left. The other focus, placed symmetrically to the right of the Sun, is empty.

Kepler's second law states that a line connecting a planet to the Sun sweeps out equal areas in equal intervals of time. Therefore, the "pie slices" shown in the illustration on the right are equal in area. Note that a planet must travel a longer distance to cover the same amount of area when closer to the Sun. Because a planet must cover the same area in an equal amount of time, it also travels faster when near the Sun.

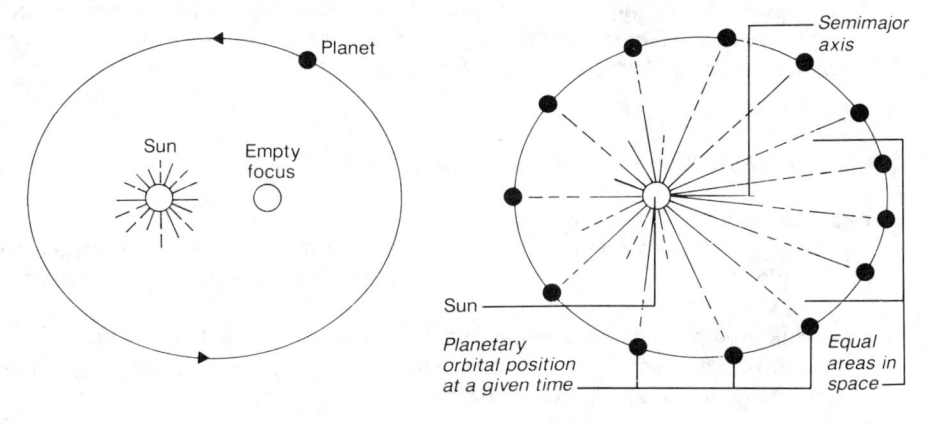

E. Galileo Galilei (1564 - 1642)
1. Galileo lived in Italy, still dominated by the Roman Catholic church; he quickly developed a reputation for challenging accepted belief systems, and devoted himself to proving Sun-centered cosmology
2. His early experiments in *mechanics,* the science of motion, overturned some of the teachings of Aristotle
   a. He discovered the concept of **inertia**, which states that an object in motion will tend to stay in motion unless a force acts to stop it
      (1) Aristotle thought that a force was always needed to maintain motion
      (2) Aristotle did not recognize that friction exerts a force that halts motion in most circumstances
   b. Galileo found that falling objects all accelerate at the same rate, regardless of weight
      (1) Aristotle stated that the rate of fall of an object depends on its weight
      (2) The famous experiment in which Galileo dropped balls of different weights from the Leaning Tower of Pisa in order to demonstrate this point probably never actually occurred, but was cited by Galileo as a thought experiment
   c. He experimented with pendulums, showing that the period of oscillation of a pendulum was constant, unaffected by the range of motion (as long as the motion was small)
3. Galileo was the first to systematically observe the nighttime sky with a telescope
   a. He discovered multitudes more stars than had been suspected; this was uncomfortable for adherents of the view that the stars were points of light attached to a rigid, crystalline sphere

b. He found craters and mountains on the Moon; this violated the notion that all celestial bodies were perfect spheres
c. He found that Jupiter has four moons orbiting it, showing clearly that some heavenly bodies do not orbit Earth
d. He discovered that both Venus and Mars undergo variations in apparent size (and phase, in the case of Venus); their motions can be explained only if they orbit the Sun
e. He showed that sunspots must be blemishes on the surface of the Sun, thus strongly suggesting that heavenly bodies can be imperfect
4. Galileo published his arguments in favor of the heliocentric theory, arousing great controversy; the Roman Catholic church censured him, and he was placed under house arrest for the last several years of his life

III. Isaac Newton and the Laws of Motion

A. General information
1. By the time of Galileo's death, his writings, along with the contemporaneous work of Kepler, had firmly established the heliocentric model despite the opposition of the Church
2. The work of Kepler and Galileo was *empirical,* meaning based on experimentation and observation; it remained for others to develop an understanding of the reasons for the observed relationships

B. Isaac Newton (1643 - 1727)
1. Newton lived in England, where he was educated in an upper-class environment
2. Newton provided the mathematical and physical framework to explain the discoveries of Kepler and Galileo
   a. In a brief period of productivity and genius (1665 - 1667), Newton made major advances in several fields
      (1) In mathematics, he invented *calculus*, a technique necessary to solve the equations representing his new discoveries in physics
      (2) He performed experiments in *optics*, leading to an understanding of many important principles of light and to the development of the first reflecting telescope
      (3) He founded the science of *mechanics*, in which the motions of bodies are studied and analyzed
   b. Newton did not publish his findings until much later (1687), in the *Philosophiae Naturalis Principia Mathematica* (commonly known as the *Principia*)
3. Newton's three laws of motion, along with his law of universal gravitation, formed the basis of an entirely new understanding of the universe
   a. Newton considered the laws of motion to be self-evident, and relegated them to an introductory preface in the *Principia*
   b. The law of universal gravitation, like the laws of motion, was applied to the universe as a whole, setting the stage for modern cosmology
   c. Newton's laws remain valid today, although they have been found subordinate to the theory of general relativity developed in the early 1900s by Einstein
      (1) Newton's laws are valid for most situations

(2) Under some circumstances involving very high velocities, very large accelerations, or very large masses, relativity must be used

C. The laws of motion
1. The law of inertia (already discovered by Galileo), also known as the first law of motion, holds that a body in a state of rest or of uniform motion remains in that state unless an unbalanced force acts on it
   a. *Inertia* is the tendency of a body to remain in its state of rest or constant motion
   b. Here *uniform motion* means motion at a constant speed in a constant direction; any deviation in either speed or direction is an *acceleration*
   c. Inertia is very closely related to *mass*, which measures the quantity of matter contained in a body
      (1) The more mass a body has, the greater the tendency to remain in a state of rest or constant motion
      (2) Mass is often confused with weight, but weight depends on the force of gravity, while mass does not; thus an astronaut weighs less on the Moon than on Earth but has the same mass in both places
   d. The law of inertia led Newton to realize that the planets must be attracted to the Sun by some force, since the planets would otherwise fly off into space in straight lines
2. The force law, more commonly known as the second law of motion, states that when a force is applied to a body, the body is accelerated in the direction of the force with an acceleration that is proportional to the force and inversely proportional to the mass of the body
   a. Mathematically, this is written as a $F = ma$, where F is the applied force, m is the mass of the body, and a is the acceleration
   b. In a friction-free environment (such as in space or on a very slippery surface), this law shows that if you push an object twice as hard one time as the other, it will accelerate twice as much
   c. Similarly, if you give equal pushes to two different objects, with one being three times more massive than the other, the less massive one will accelerate three times as much
3. The law of action and reaction, also known as the third law of motion, states that forces always occur in pairs; for every force there is an equal and opposite force
   a. This is easy to see in some situations where motion occurs
      (1) For example, when a shotgun is fired, the gun "kicks" as the pellets are accelerated through the barrel
      (2) This is the principle on which a rocket works; gas is expelled explosively through a nozzle, accelerating the rocket in the opposite direction
   b. The "action" and "reaction" are not so obvious in static situations, where no motion occurs
      (1) For example, a chair exerts a downward force on the floor that is balanced by an upward force exerted by the floor on the chair
      (2) Inside a star, the inward force of gravity is balanced everywhere by an outward pressure force

D. The law of universal gravitation
1. The fact that the Moon orbits Earth and the planets orbit the Sun requires that a force attract these bodies to each other
   a. The law of inertia led Newton to realize that the planets and the Moon would fly off in straight lines if there were no force
   b. The curving path of an orbiting body is a combination of its forward momentum and inward acceleration caused by this force
2. Newton was able to determine the nature of this force, which is called gravity
3. Newton's *law of universal gravitation* states that every pair of objects in the universe attracts each other with a force that is proportional to the product of their masses and inversely proportional to the square of the distance between them
   a. Mathematically, this is written as $F = Gm_1m_2/r^2$, where F is the force, G is the gravitational constant, m_1 and m_2 are the masses of the two bodies, and r is the distance between them
      (1) The value of the gravitational constant is 6.67×10^{-11} N·m²/kg²
      (2) One N (or newton) equals 1 kg·m/sec²
   b. It is possible to perform thought examples as an aide in understanding the form of this law
      (1) Consider the Earth-Moon system, and suppose that the mass of the Moon is doubled; thus, the force between them is doubled
      (2) Consider that the mass of the Moon is tripled while the mass of Earth is increased fourfold and the distance between them is doubled; then the force between them is tripled
4. The surface gravity of a planet can be easily derived using the law of gravitation
   a. If m is the mass of an object and M is the mass of the planet and R is the planet's radius, then the gravitational force between the object and the planet is $F = GmM/R^2$
   b. Using the second law of motion, we can state any force in terms of mass times acceleration, so if g is the acceleration of gravity at the surface of the planet, we have $F = mg$
   c. Setting the two expressions for the force equal to each other, we have $mg = GmM/R^2$
   d. The m's cancel out, and we have $g = GM/R^2$; this expression for the gravitational acceleration g does not depend on the mass of the object, but only on the mass and radius of the planet
5. Using Newton's laws and the concept of energy, it is possible to derive the *escape speed* from a planet; this is the upward speed required in order for a body to escape the planet's gravitational field
   a. Energy is defined as the ability to do work
      (1) **Kinetic energy** is energy of motion; the expression for the kinetic energy of a body of mass m moving at velocity v is $E = mv^2/2$
      (1) **Potential energy** is stored energy; that is, energy that must be released in order to do work
      (2) A body suspended in a gravitational field has potential energy because it will fall and create kinetic energy if released
      (3) The expression for gravitational potential energy is $E = GmM/R$, where m is the mass of the object, M is the mass of the planet, and R is the radius of the planet

b. The escape speed is found by setting the gravitational potential energy equal to the kinetic energy, and solving for v
      (1) Setting the two terms equal yields $mv^2/2 = GmM/R$
      (2) Solving for v yields:
      $$\sqrt{\frac{2GM}{R}}$$
      (3) As an example, we find that the escape speed for Earth is:
      $$v = \sqrt{\frac{(2)(6.67 \times 10^{-11} \text{ N} \cdot \text{m}^2/\text{kg}^2)(5.974 \times 10^{24} \text{ kg})}{(6.378 \times 10^6 \text{ km})}}$$
      $$= 11{,}187 \text{ m/sec or } 11.2 \text{ km/sec}^2$$

E. Newton's derivation of Kepler's laws of planetary motion
   1. Newton's laws permit a theoretical, mathematical description of orbits, improving upon the empirical descriptions of Kepler
   2. Newton found that when two bodies are in orbit, each orbits the **center of mass**
      a. The center of mass is the point in space between the bodies where the product of the mass times the distance from this point is equal for the two (see *Center of Mass*, page 26)
      b. For example, if one body is twice as massive as the other, the more massive one is half as far from the center of mass
   3. Kepler's first law is modified to state that each planet orbits the center of mass of the planet-Sun pair in an ellipse, with the center of mass at one focus
      a. Each planet is very much less massive than the Sun, so the center of mass in each case is near the center of the Sun
      b. Thus Kepler was unaware of the distinction between the location of the Sun and the center of mass
      c. For orbiting bodies that are more nearly equal in mass, such as a double star system, the center of mass is closer to the midpoint between the two bodies
   4. Kepler's second law, as stated by Newton, says that the *angular momentum* in a system is constant
      a. Angular momentum is the product of a body's mass, its distance from the center of mass, and the component of its orbital motion that is perpendicular to the line from the center of mass to the body
      b. For a circular orbit, angular momentum is the product mvr, where m is the mass of the orbiting body, v is its orbital speed, and r is its orbital radius
      c. This gives a mathematical explanation for the fact that a planet moves faster when it is closer to the center of mass, and slower when it is farther away
   5. Newton's form of Kepler's third law relates the masses of the two bodies to their orbital size and period; it states that if m_1 and m_2 are the two masses, then the law becomes $(m_1 + m_2)P^2 = a^3$, if the masses are in units of solar mass, the period (P) in years, and the semimajor axis (a) in astronomical units
      a. Kepler did not recognize the dependence on mass because for each Sun-planet pair, the total mass is approximately equal to the Sun's mass alone
      b. This form of Kepler's third law is very useful for determining the masses of distant bodies because it is often possible to determine P and a from observation, and then solve the equation for the sum $(m_1 + m_2)$

Center of Mass

In the upper diagram, star A has twice the mass of star B, and the two stars orbit around a center of mass that is one-third the way between the centers of the two stars. In the bottom sketch, star B is a tenth the mass of star A; hence the center of mass is quite close to star A. For any Sun-planet pair in our solar system, the center of mass is close to the Sun's center because the Sun is much more massive. Consequently, the Sun's orbital motion is minimal.

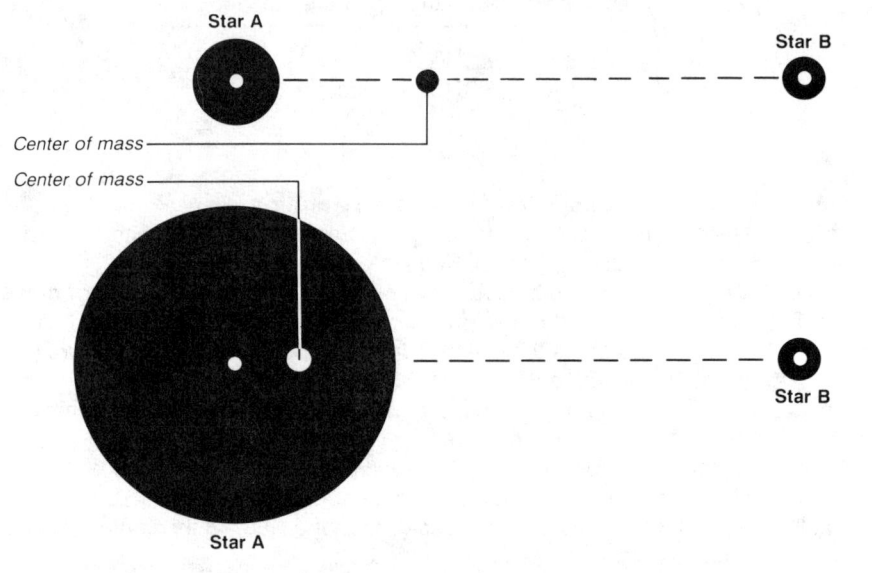

    (1) In order to find the two masses individually, more information is needed (see Chapter 12, Measuring the Stars)
    (2) Use of Kepler's third law in this form is extremely important in astronomy because it is the only method available for finding masses of distant bodies
  c. If standard physics units are used, Newton's form of Kepler's third law becomes $(m_1 + m_2)P^2 = (4\pi^{2/G})a^3$, where π and G are constants
    (1) As an example, the mass of Jupiter can be found by measuring the orbital period and semimajor axis for one of its satellites; using the moon Callisto's period of 16.689 days = 1.442×10^6 sec and semimajor axis of 1.88×10^9 m yields a total mass of Callisto plus Jupiter of $(m_1 + m_2) = (4\pi^{2/G}(a^3)/P^2) = 1.89 \times 10^{27}$ kg; since Callisto is very much smaller than Jupiter, we can neglect Callisto and conclude that this is the mass of Jupiter
    (2) In most applications, especially involving stellar masses, it is more convenient to keep the masses in astronomical units of solar mass and the periods in units of years, so that the simpler form of Kepler's third law can be used

F. Tides
1. *Tides* are created by the difference in gravitational forces acting on opposing sides of a body
   a. Because the gravitational force between two bodies decreases in relation to the square of their separation, the force on the near side is greater than the force on the far side
   b. The difference between forces acting on opposite sides is greater the closer the two bodies are
2. The difference in force acting on opposite sides of a body is called a **tidal force,** or **differential gravitational force**
3. A tidal force tends to stretch a body along the direction of the nearby body that is exerting the force
   a. This stretching occurs because the far side of the body is susceptible to less gravitational force than its center, while the near side feels a stronger gravitational force than the center; thus both sides are pulled away from the center
   b. On Earth this causes the oceans to rise into global ridges on the side facing the Moon and on the opposite side
   c. As Earth spins, these two ridges of water travel over its surface, causing two high and two low tides per day at any given location
   d. In a few areas on Earth, such as the Gulf of Mexico, local effects of the water basin act to modify tidal forces, and these areas have only one high tide and one low tide per day
4. Earth also exerts a tidal force on the Moon, causing it to have an elongated shape even though it does not have oceans
   a. When the Moon was younger, it spun faster than it does now, and this tidal bulge travelled over its surface
   b. This created internal friction and heat dissipation in the Moon
   c. Loss of energy gradually slowed the Moon's spin, until it reached the point where its rotation period equaled its orbital period and it could keep the same side toward Earth at all times
5. Consequently, tidal forces have locked the Moon into **synchronous rotation,** meaning that its orbital and spin periods are equal
   a. Most of the satellites in the solar system are in synchronous rotation, due to tidal forces exerted by their parent planets
   b. The Moon's tidal force on Earth has similarly slowed Earth's rotation, but not by much; it will take many billions of years for Earth to be locked into synchronous rotation with the Moon

Study Activities
1. Summarize the contributions of the Babylonian culture that directly affect us today.
2. Compare and contrast the Greek explanation of retrograde motion with the modern explanation.
3. Explain why the Sun-centered cosmology of Aristarchus was not accepted at the time he proposed it.
4. Summarize Copernicus's reasons for adopting the heliocentric model of the universe.

5. Explain why Brahe's practice of making multiple measurements of planetary positions, at every opportunity, was important for the subsequent understanding of planetary motions.
6. Explain why it was fortunate that Kepler chose to work with the data on Mars as he developed his laws of planetary motion.
7. Summarize the ways in which the deductions of Galileo contradicted the teachings of Plato and Aristotle.
8. Explain how Newton realized that there must be a force attracting the planets toward the Sun; explain how he deduced the nature of this force (that is, how he concluded that its strength is inversely proportional to the square of the distance between two bodies).
9. Calculate the mass of the Earth-Moon system, using Newton's form of Kepler's third law, the orbital period of the Moon, and the length of the Moon's semimajor axis.
10. Calculate the surface gravity and the escape speed for the planet Jupiter.

4

Light and the Atom

Objectives

After studying this chapter, the reader should be able to:
- Describe the many forms of electromagnetic radiation.
- Discuss the characteristics of the photon.
- Apply the laws of continuous radiation.
- Describe the formation of emission and absorption lines by electrons in atoms and ions.
- Calculate radial velocities using the Doppler effect.

I. Electromagnetic Radiation

A. General information
 1. Light is a form of ***electromagnetic radiation;*** it consists of oscillating electric and magnetic fields
 2. Electromagnetic radiation displays properties of waves and particles
    a. Wavelike properties include refraction, diffraction, and interference (see part C. below)
    b. Particle-like properties include the ability to travel in a vacuum and to carry energy only in discrete quanta (see part C. below)
 3. The particle vs. wave issue was not settled until the early 20th century
    a. Newton, in the late 17th century, favored the particle interpretation
    b. His contemporary, the Dutch scientist Christiaan Huygens, argued for the wave hypothesis
    c. The modern work of Niels Bohr, Max Planck, Albert Einstein, and others helped prove that electromagnetic radiation has qualities of both waves and particles (see part C. below)
 4. Electromagnetic radiation can be visible or invisible, depending on its *wavelength* (distance between crests of successive waves)
 5. Electromagnetic radiation is important to astronomy because observations of radiation emitted or reflected by celestial objects provide almost the only information that can be obtained about distant objects

B. The electromagnetic spectrum
 1. Light has wavelike properties, including wavelength; the human eye sees different wavelengths of visible light as different colors
 2. The wavelength (λ) of a wave is the distance from one wave crest to the next

      a. A unit commonly used by astronomers is the *angstrom* unit (Å); one angstrom equals 1×10^{-10} m
      b. Many physicists and astronomers are now using the *nanometer;* one nanometer equals 1×10^{-9} m
  3. A *spectrum* is a display or graph showing the variation of intensity of a light source according to wavelength; the full range of all possible wavelengths (including invisible ones) is called the **electromagnetic spectrum**
  4. The *frequency (f)* is the number of waves per second passing the observer
      a. If c is the speed of light, then $\lambda = c/f$
      b. The standard unit of frequency is the *hertz;* one hertz equals one wave per second (or one cycle per second when discussing oscillations, such as in household electrical current or a vibrating spring)
  5. The electromagnetic spectrum consists of several types of radiation, all identical except for wavelength and frequency, and each one merging continuously into the next
      a. Gamma rays include all wavelengths shorter than 1×10^{-10} m (shorter than 1 Å or 0.1 nm)
      b. X-rays range in wavelength from 1×10^{-10} m to 1×10^{-8} m (1 to 100 Å or 0.1 to 10 nm)
      c. Ultraviolet radiation has wavelengths between 1×10^{-8} m and 4×10^{-7} m (100 to 4,000 Å or 10 to 400 nm)
      d. Visible light, which comprises only a tiny fraction of the electromagnetic spectrum, lies between 4×10^{-7} m for violet light and 7×10^{-7} m for red light (4,000 to 7,000 Å or 400 to 700 nm)
      e. Infrared radiation ranges in wavelength from 7×10^{-7} m to 1×10^{-4} m (7,000 to 1,000,000 Å or 700 to 100,000 nm)
        (1) Astronomers studying infrared radiation commonly use the *micron* (μ), equal to 1×10^{-6} m
        (2) In this unit, infrared radiation ranges in wavelength from 0.7 to 100 μ
      f. Radio waves have wavelengths longer than 1×10^{-4} m (longer than 100 m)
        (1) The shortest-wavelength radio radiation, from 100 μ to about 10 cm, commonly is called microwave radiation
        (2) There is no rigid long-wavelength limit to the radio portion of the spectrum, but few natural processes can produce radiation with wavelengths longer than a few km
  6. The universe and the objects in it emit and absorb electromagnetic radiation throughout the spectrum; thus astronomers need telescopes for all types of radiation, not just visible light

C. Particles or waves?
  1. Electromagnetic radiation has wavelike properties
      a. It displays *refraction,* meaning that its speed and direction change when it passes through a boundary between one medium and another
      b. It exhibits *diffraction,* meaning that its path bends as it passes the edge of an obstruction
      c. It undergoes *interference,* meaning that when electromagnetic waves meet, they can combine constructively if their crests coincide or destructively if the crests from one wave coincide with the troughs of the other
  2. Electromagnetic radiation also has particle-like properties

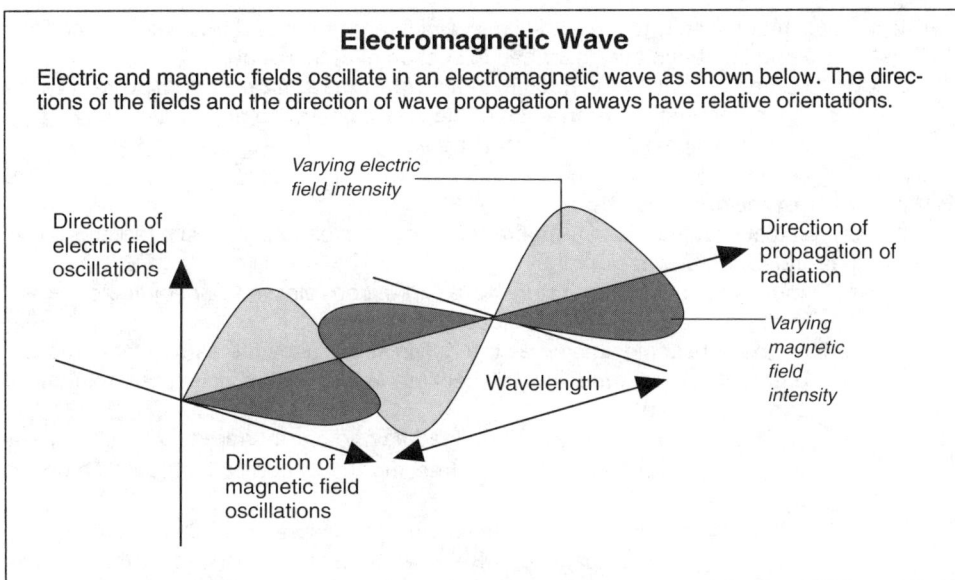

Electromagnetic Wave

Electric and magnetic fields oscillate in an electromagnetic wave as shown below. The directions of the fields and the direction of wave propagation always have relative orientations.

    a. It carries energy only in discrete quantities called *quanta*
    b. It can travel through a vacuum
3. The wavelike and particle-like properties are combined in the modern concept of the ***photon***
    a. A photon can be viewed as a packet of electromagnetic waves that has a fixed energy and travels like a particle
    b. The energy (E) of a photon is proportional to its frequency (f), or inversely proportional to its wavelength; this relationship can be expressed mathematically as $E = hf$ (or $E = hc/\lambda$), where h is the Planck constant (6.62068×10^{-34} joule·sec) and c is the speed of light (2.9979×10^8 m/sec)
    c. The electric and magnetic fields in a photon oscillate in phase with each other (that is, their peaks and troughs coincide), but in planes that are at right angles to each other (for a schematic representation of an electromagnetic wave, see *Electromagnetic Wave*)
    d. Radiation (that is, a stream of photons) is *polarized* if the orientation of the electric and magnetic fields is not random — one orientation tends to be more common than others
        (1) Polarization can be caused by the nature of the emitting source of radiation
        (2) Polarization also can be caused by the nature of the medium through which the radiation travels

II. Continuous Radiation

A. General information
    1. In *continuous radiation,* most natural sources of radiation are emitted over a broad range of wavelengths

2. Many sources also have spectral lines, which are bright or dark lines at specific wavelengths; these are described in section III of this chapter
3. Most sources of continuous radiation emit **thermal radiation** or **blackbody radiation,** meaning that the properties of the emission are determined entirely by the surface temperature of the emitting source

B. Properties of thermal radiation
 1. Every object with a surface temperature above absolute zero emits radiation over a broad range of wavelengths
    a. In most sciences, temperatures are expressed relative to *absolute zero*, the temperature at which all molecular motions stop
    b. This absolute scale, known as the Kelvin scale, uses the same unit of measurement—the degree—as the centigrade, or Celsius, scale, but not the same zero point
       (1) The centigrade degree is defined as one one-hundredth of the temperature difference between the freezing (0° C) and boiling (100° C) points of water
       (2) On the centigrade scale, absolute zero occurs at $-273°$ C
       (3) Hence, on the absolute or Kelvin scale, zero is absolute zero (0 K = $-273°$ C) and the freezing point of water is 273 K; normal room temperature is about 295 K
       (4) The unit of temperature on this absolute scale is called the *kelvin,* not the Kelvin degree; hence we say that water boils at a temperature of 373 kelvins (373 K), not 373 degrees Kelvin (373° K)
 2. The *Planck function* is a mathematical expression that gives the spectrum of a thermal source (as a function of wavelength or frequency) based solely on the surface temperature of the source
 3. Several simple laws describing the properties of thermal radiation have been discovered experimentally and also can be derived from the Planck function
    a. *Wien's law* states that the wavelength of peak emission intensity (λ_{max}) of a source is inversely proportional to the surface temperature of the source
       (1) Mathematically, this is expressed as $\lambda_{max} = W/T$, where W is a mathematical constant (0.0029 m · degree) and T is the surface temperature in kelvins
       (2) For example, the Sun's surface temperature is approximately 5,780 K, and its wavelength of maximum emission (assuming it is a perfect thermal emitter) is $\lambda_{max} = 0.0029$ m · degree/5,780 k = 5.017×10^{-7} cm
    b. The *Stefan-Boltzmann law* states that the total energy emitted per square meter per second (that is, the power) by a thermal source is proportional to the fourth power of the surface temperature
       (1) Mathematically this is written as $P = \sigma T4$, where σ is the *Stefan-Boltzmann constant* (5.67×10^{-8} W/m^{2}k^4) and the W in the Stefan-Boltzmann constant refers to watts
       (2) For example, the power emitted by each square meter of the Sun's surface is $P = (5.67 \times 10^{-8}$ W/m^{2}k$^4)(5,780$ k$)^4 = 6.33 \times 10^7$ W/m^2
    c. The **luminosity (L)** of a thermal source is the total energy emitted per second over the entire surface area

(1) The luminosity is equal to the energy emitted per square meter per second (from Stefan's law) multiplied by the surface area in square meters
(2) For a spherical object such as a star, this yields $L = 4\pi R^2 \sigma T^4$, where R is the radius and T is the surface temperature in kelvins (this expression is sometimes known as the Stefan-Boltzmann law)
(3) For example, if we multiply the Sun's power of 6.33×10^7 W/m^2 times $4\pi R^2$ (where R is the Sun's radius of 7×10^8 m), we find that the Sun's luminosity is $L = 3.9 \times 10^{26}$ W (or 3.9×10^{24} 100-watt light bulbs)
4. For any source, thermal or not, the observed brightness varies inversely with the distance between observer and source
   a. For example, if two stars have equal luminosity and one is five times farther away than the other, the more distant one will appear $(1/5)^2 = 1/25$ as bright
   b. Astronomers always take into account the dependence of apparent brightness when comparing the intrinsic properties of objects such as stars

III. Spectral Lines and the Atom

A. General information
1. The spectra of many light sources have either bright or dark **spectral lines** at specific wavelengths
   a. The bright lines, called **emission lines,** are wavelengths where the source emits more radiation than at adjacent wavelengths
   b. The dark lines, called **absorption lines,** are wavelengths where the source emits less radiation than at adjacent wavelengths
2. Spectral lines contain an immense amount of information about a source of radiation and about the medium through which the radiation travels
   a. Factors such as the chemical composition of a source, its temperature, surface density and pressure, the strength of the magnetic field, and its motion can all be derived from analysis of spectral lines
   b. Thus, the analysis of spectra, called spectroscopy, is the most common form of observational research
3. In order to understand and interpret spectral lines, it is necessary to understand the structure of the atom

B. Observed properties of spectral lines
1. Each chemical element (that is, each kind of atom) has its own unique set of spectral lines, which can be used to identify specific elements (or ions or molecules); wavelengths for a given element are the same whether the lines are emission or absorption ones
2. Experiments in the mid 1800s revealed some systematic behavior of spectral lines, known as *Kirchhoff's rules*
   a. A hot solid or a dense, hot gas produces a continuous spectrum (without spectral lines)
   b. Emission lines are produced by rarefied, hot gases, such as those resulting from flames
   c. Absorption lines are produced when a cool gas lies in front of a hotter source

C. The atom and formation of spectral lines
1. An atom consists of a nucleus and a cloud of electrons that orbit the nucleus
   a. The nucleus contains **protons**, which carry positive electrical charges, and **neutrons**, which have no electrical charge
   b. The **electrons** carry negative electrical charges and normally exist in equal numbers to the protons in the nucleus, so that the atom is electrically neutral
2. Spectral lines are formed by the electrons orbiting the nucleus (see *Formation of Spectral Lines*)
   a. Absorption lines are formed when electrons absorb photons of light
   b. Emission lines are formed when electrons emit photons of light
   c. Electrons can absorb or emit light only at specific wavelengths because the electrons can gain or lose energy only in specific amounts called *quanta*
      (1) This is because electrons orbiting a nucleus can exist only in specific, fixed energy states, and can gain or lose energy only in quantities that correspond to the differences in energy between energy states
      (2) The energy and wavelength of a photon are related (see section I., part D. above), so if an electron can gain or lose only specific quantities of energy, this means it can gain or emit photons only at specific wavelengths
3. The wavelengths of spectral lines formed by an atom depend on the specific structure of electron energy states for that atom
   a. Each kind of atom has its own unique set of electron energy states; therefore, each element has its own unique set of spectral line wavelengths
   b. The wavelengths are the same for absorption and emission lines because the electron energy structure is fixed for each kind of atom; the only difference is whether the electrons are gaining energy by absorbing photons of appropriate wavelength, or losing energy by emitting photons of appropriate wavelength
4. Whether emission or absorption occurs depends on physical conditions
   a. Electrons can change energy states through collisions between atoms, in which energy is transferred from one atom to the other
      (1) The temperature of a gas plays a role because temperature is linked to the average speed of the individual atoms; speed, in turn, determines how much energy can be transferred from atom to atom via collisions
      (2) Density also plays a role, because the number of atoms per unit of volume governs the frequency of collisions
   b. In a cool, thin gas, collisions are rare and low-energy, so electrons tend to be in low-energy states from which they cannot emit but can absorb; thus a cool gas forms absorption lines if a source of continuous radiation is viewed through it (this explains Kirchhoff's *third rule* as stated above)
   c. In a hot, thin gas, collisions occur at substantial energies; thus, at any moment, some electrons are in high-energy states from which they can drop to lower ones, emitting photons and creating emission lines (this explains Kirchhoff's *second rule*)
   d. In a hot, dense gas, collisions between atoms are both frequent and high-energy, and the collisions distort electron energy levels so that the discrete energy levels are blended together; thus, such a gas emits continuously (this explains Kirchhoff's *first rule*)

Formation of Spectral Lines

Several orbits are possible for an electron; each orbit represents a different energy state. Absorption lines are formed if an electron absorbs exactly the amount of energy needed to jump to a higher orbit; the energy difference corresponds to the wavelength of the photon absorbed; this is fixed for any given kind of atom. Conversely, an electron emits a photon when it drops from one orbit to a lower one; the photon's wavelength corresponds to the energy difference between orbits. This is how emission lines form. Note that an electron can gain enough energy to escape from the atom altogether, a process called ionization.

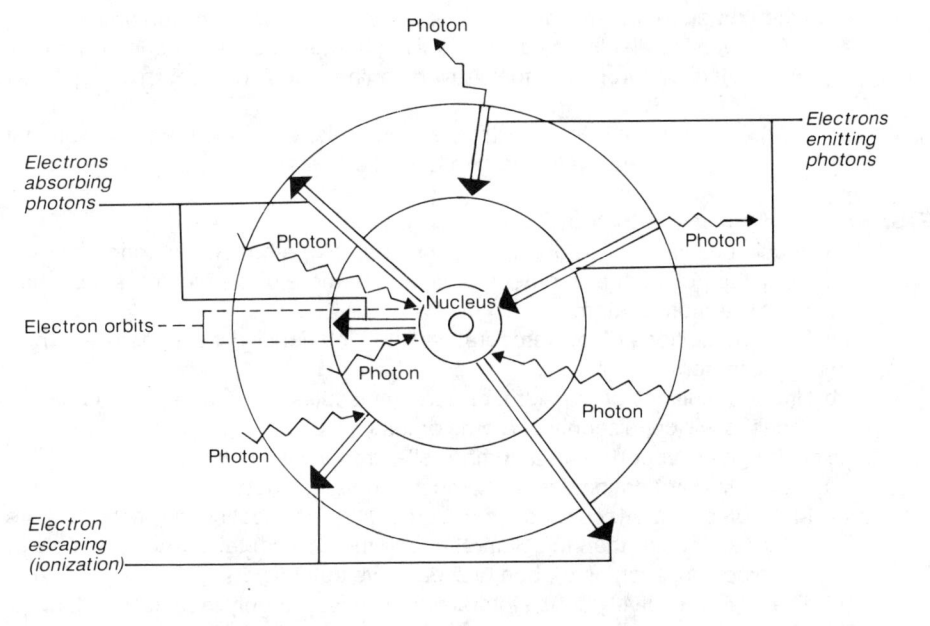

D. Excitation and ionization
1. Because collisions affect the spectral line formation of a gas and are themselves governed by physical conditions in the gas, it is possible to determine physical conditions through analysis of spectral lines
2. Any process that raises electrons to energy levels above the lowest possible one is called **excitation**
    a. Excitation can occur when an electron absorbs a photon and jumps to a higher energy level
    b. Excitation also can occur when an electron gains energy from a collision with another particle and jumps to a higher energy level
        (1) Scientists can determine the degree of excitation of a gas, because the absorption lines formed depend on the energy levels of the electrons
        (2) Therefore, the density of a gas can be determined from its absorption lines; the degree of excitation depends on the density
3. Any process in which electrons are freed from their nucleus is called **ionization**
    a. Ionization can occur if an electron absorbs a photon having enough energy to free the electron; this is called *photoionization*
    b. Ionization also can occur if an electron gains enough energy from a collision with another particle to be freed from its nucleus

(1) The energy of particle collisions depends on the temperature of a gas; the hotter the gas, the faster the particles move and the higher the energy of collisions
(2) The degree of ionization thus depends on gas temperature; the hotter the gas, the more highly ionized it is (that is, the greater the fraction of the electrons that have been freed from their nuclei)
(3) Therefore, the temperature of a gas can be determined from its spectral lines, because each ion has its own unique set of spectral-line wavelengths
   c. When an atom has lost one or more of its electrons, it is called an *ion*
      (1) Because electrons carry negative charges, the loss of one or more electrons from an atom leaves behind a positively charged ion, known as a *cation*
      (2) An atom can gain an extra electron under some circumstances, resulting in a negatively charged ion called an *anion*

E. Spectral line formation by molecules

1. A molecule consists of two or more atoms bound together and sharing one or more of their electrons; in other words, one or more of the electrons orbit the entire multi-atom system
   a. Under conditions of low temperature and high density, most gases are in molecular form
   b. Under conditions of high temperature, molecules are destroyed, and most gases will consist only of atoms or ions
2. Molecular gas is important in a number of astronomical situations
   a. Molecules predominate in the atmospheres of planets
   b. Molecules also are common in the outer layers of cool stars (primarily stars that are cooler than the Sun, although the Sun contains trace amounts of molecules, such as carbon dioxide, in its outer layers)
   c. The gas in relatively dense interstellar clouds is largely in molecular form
3. The electrons orbiting a molecule have fixed orbits just as those orbiting atoms or ions, and they form spectral lines in the same manner
4. In addition, molecules can form absorption and emission lines by other mechanisms
   a. A molecule has fixed energy states corresponding to vibrations of atoms within the molecule; changes between vibrational states can form emission or absorption lines (generally in the infrared or microwave portion of the spectrum)
   b. A molecule also has fixed energy states corresponding to differences in the rates of rotation of the molecule, and changes in rotational energy can form emission or absorption lines (generally in the microwave portion of the spectrum)
5. Consequently, a molecular gas has a complex spectrum of emission or absorption lines

F. The Doppler effect

1. The **Doppler effect** is the shift of wavelength (or frequency) that occurs when there is motion along the line of sight between the source and the observer of waves

a. The effect was originally discovered and analyzed in connection with sound waves
   b. The Doppler effect also occurs in light, although some differences between the effect in light and in sound exist
2. If a source and observer are approaching each other, all spectral lines are shifted toward shorter wavelengths; this is called a **blueshift**
3. If a source and observer are moving away from each other, all spectral lines are shifted toward longer wavelengths; this is called a **redshift**
4. The relative speed along the line of sight of source and observer, called the *radial velocity (v)*, is determined by the formula $v = c \times (\Delta\lambda/\lambda)$, where c is the speed of light, λ is the laboratory (rest) wavelength of the spectral line, and $\Delta\lambda$ is the shift of the line's wavelength (negative for approach; positive for recession)
   a. For example, suppose a star is observed in which a spectral line known to have a laboratory wavelength of 6.563×10^{-7} m is observed to have a wavelength of 6.565×10^{-7} m; then the shift is 2×10^{-10} m and the radial velocity is $v = c(\Delta\lambda/\lambda) = (2 \times 10^{-10}/6.563 \times 10^{-7})(2.9979 \times 10^8$ m/sec$) =$ 9,008 m/sec (or approximately 9.0 km/sec)
   b. The radial velocity consists only of the portion of the relative motion that is directed along the line of sight between source and observer; motion perpendicular to the line of sight causes no Doppler shift

Study Activities

1. Summarize the evidence favoring the particle nature of light. What is the evidence supporting the viewpoint that light consists of waves?
2. Explain how the modern photon theory incorporates both the wave and particle nature of light.
3. Using the expressions relating energy, wavelength, and frequency of a photon, calculate the following:
   (a) the frequency of a photon of visible light, having a wavelength of 5.5×10^{-7} m
   (b) the wavelength of your favorite radio station (typically, the frequency is given; be sure to get the units correctly—a *kilohertz* is 1,000 Hz and a *megahertz* is 1,000,000 Hz)
   (c) the energy of the same photon of visible light referred to in (a)
   (d) the energy of an X-ray photon having a wavelength of 50Å
   (e) the energy of a radio photon having a wavelength of 10 m.
4. Use Wien's law to determine at what wavelengths the following are emitting most of their radiation: the Sun, with a surface temperature of 5,780 K; a neutron star, with a surface temperature of 2,000,000 K; your body, with a surface temperature of about 300 K. Can your eyes see the radiation from each of these objects? Explain.
5. Calculate the power emitted by one square meter of surface on a star having a surface temperature of 20,000 K. Do the same for a star having a surface temperature of 2,000 K.

6. Now calculate the luminosities of both stars in question 5, assuming that the 20,000 K star has a radius of 5×10^9 m and the 2,000 K star has a radius of 2×10^8 m. Compare your answers to the luminosity of the Sun, whose luminosity is approximately 4×10^{26} watts.

7. Suppose the star Imagina, identical to the Sun in luminosity, is seen at a distance of 1 million AU. How much dimmer than the Sun will Imagina appear? Now suppose another Sun-like star, Hypothetica, is found to be 100 times brighter than Imagina; how much closer to us is Hypothetica, compared to Imagina?

8. Explain, using your own words, the fact that each kind of atom or ion has its own unique set of wavelengths at which it forms spectral lines.

9. Explain how the excitation of a gas is related to its density and how ionization is related to its temperature. How do excitation and ionization affect the spectrum of a gas?

10. Find the radial velocity of a star in whose spectrum the line normally seen at 6.6563×10^{-7} m is observed at a wavelength of 6.6558×10^{-7} m.

5

Telescopes

Objectives

After studying this chapter, the reader should be able to:
- Explain the many advantages telescopes offer over naked-eye observations.
- Discuss the contrasts between refracting and reflecting telescopes.
- Describe the basic design of a large modern telescope.
- Summarize the characteristics of a good observatory site.
- Describe the new techniques being developed to build larger telescopes.
- Summarize the design and operating principles of radio telescopes.
- Describe the design of telescopes for gamma-ray, X-ray, ultraviolet, and infrared astronomy.

I. Telescopes for Visible Light

A. General information
 1. Until the early 17th century, all astronomical observations were done without telescopes
    a. Observations consisted almost entirely of positional measurements of objects bright enough to be seen by the naked eye
    b. In 1609, Galileo was the first person to use a telescope to study the skies
 2. The earliest telescopes were **refractors,** using lenses to bring light to a focus
 3. Newton invented the first **reflector,** using mirrors to focus light; today, most telescopes are reflectors
 4. Until the 1930s, when the first radio telescope was built, all telescopes were designed for visible-wavelength light
    a. Radio astronomy was not actively pursued until the 1950s
    b. In the 1960s and 1970s, instruments measuring ultraviolet, X-ray, and infrared radiation began to be developed
    c. Today astronomers can obtain data on celestial objects at virtually any wavelength in the electromagnetic spectrum

B. The advantages of telescopes
 1. Telescopes collect light over a much larger area and concentrate its energy into a much smaller area than the human eye, allowing observations of much fainter objects
    a. The *light-collecting power* of a telescope is proportional to the collecting area, which is the square of the diameter of the telescope

b. The largest telescope currently in operation (the 10-m Keck telescope) has light-collecting power approximately one trillion times greater than the human eye
2. Telescopes gain an additional advantage over the human eye by collecting light from a source over long periods of time
   a. The human eye records light only for a small fraction of a second; a telescope employing a camera or other device can record light over several hours
   b. In order to observe an object over an extended period, a telescope must have a drive mechanism to keep the telescope pointed at the source despite Earth's rotation
3. Telescopes also provide sharper images than the human eye
   a. The **resolving power,** or **resolution,** of telescopes is a measure of the sharpness of an image and is defined by the smallest angular separation that can be discerned (see *Resolving Power of a Telescope*)
   b. The resolving power of the human eye is approximately 2 arcminutes (2′); this would be the angular size of the smallest detail the eye could see by itself
   c. The resolving power of a telescope is proportional to the wavelength being observed and inversely proportional to the diameter of the telescope
      (1) Thus a 10-cm telescope (about 100 times larger in diameter than the pupil of the eye) has a resolution of 1/100 of two arcminutes, or about 1 arcsecond (1″)
      (2) In principle even larger telescopes have even better resolution, but Earth's atmosphere causes blurriness, which normally limits resolution to about 1 arcsecond, regardless of telescope size
4. An additional advantage of telescopes is that instruments can be attached to record and analyze the light

C. Principles of telescope design
1. Refractors, or refracting telescopes, form images using lenses
   a. A lens forms an image because light rays are bent (refracted) as they enter the lens
      (1) This bending occurs because the light changes speed as it passes from one medium (such as air) into another (such as glass)
      (2) The degree of bending depends on the nature of the media (which governs the amount by which the speed of light is changed) and on the angle of incidence of the light relative to the lens surface
      (3) If properly shaped, a lens will bring all rays that pass through it to the same focal point
   b. A main lens called the *objective lens* defines the light-collecting power of the telescope; at least one additional lens is needed to magnify the image
   c. A disadvantage of the refractor is that the lens can be supported only at its edges because of its curved shape; this limits the size of the lens to about one meter
      (1) Larger lenses will sag under their own weight, thus distorting images
      (2) The largest refractor is the 1.2-m telescope at the Yerkes Observatory in Wisconsin
   d. Another disadvantage is that a lens tends to separate or disperse the colors because different wavelengths of light refract at different angles; the effect

Resolving Power of a Telescope

The resolving power, or resolution, of a telescope is defined as the smallest angular distance between two objects in the sky that can be seen as separate objects. In other words, it is the smallest detail that can be discerned in angular units.

The property of light that ultimately limits the resolution of a telescope is *diffraction*; light waves that pass near the edge of the telescope opening are slightly bent in direction and interfere with nearby waves that enter in a straight line. The interference causes blurriness at an angular scale that is related to the ratio of the observed wavelength to the diameter of the telescope. For a circular opening, the *diffraction limit*, the ultimate resolution possible, is determined by $\theta = 250,000(\lambda/D)$, where θ is the resolution in arseconds, λ is the wavelength being observed, and D is the diameter of the telescope opening.

The human eye has an opening (the pupil) about 1 mm (0.1 cm) in diameter; thus, for visible light (typical wavelength of 5.5×10^{-5} cm), the resolution equals $(250,000)(5.5 \times 10^{-5}$ cm/0.1 cm$) = 138$ arcseconds or 2.3 arcminutes. For a small telescope (about 4 inches or 10 cm) at the same wavelength, the resolution is $(250,000)(5.5 \times 10^{-5}$ cm/10 cm$) = 1.4$ arcseconds. This is a very small angle; an angle of 1 arcsecond corresponds to the angular diameter of a dime seen from a distance of about 2 miles. For a large telescope (about 4 m), the resolution would be much smaller, but Earth's atmosphere usually limits it to about 1 arcsecond or only slightly better. The Hubble Space telescope (HST), with a diameter of 2.4 m (240 cm), has a resolution for visible light equal to $(250,000)(5.5 \times 10^{-5}$ cm/240 cm$) = 0.057$ arcsecond. One of the goals in building and flying the HST was to achieve resolution that is superior to what can be obtained using ground-based telescopes.

It is instructive to compute the resolution for wavelengths other than visible light. For example, calculate the resolution for a radio telescope of 100-m aperture, designed to observe at a wavelength of 21 cm. This will give you an idea why interferometry is especially important for radio telescopes.

---

is known as *chromatic aberration,* which can be corrected by using layered lenses with different refractive properties and shapes
2. Reflectors, or reflecting telescopes, bring light to a focus using mirrors
   a. Reflectors have several advantages over refractors
      (1) A mirror can be supported from behind instead of only at the edges; this allows much larger telescopes that use mirrors to be built
      (2) Only one surface of a mirror has to be shaped and polished to a high degree of accuracy, whereas a lens has two surfaces that must be shaped and polished and the interior also must be flawless
      (3) When light reflects, it does so independently of wavelength, so all colors focus at a common point without chromatic aberration
   b. The principal mirror in a reflecting telescope, called the *primary mirror*, is concave, so light striking it at any point is brought to a common focus
      (1) A *parabola* accomplishes this best for the normal circumstance where parallel rays of light enter the telescope along its axis
      (2) The area of the primary mirror, which is equal to πr^2 (where r stands for its radius), defines the light-collecting power of the telescope
   c. The image formed by the primary mirror lies in front of it, where viewing is inconvenient or impossible
   d. A *secondary mirror* is used to deflect the image to a point outside the telescope

(1) The secondary mirror blocks a small fraction of the incoming light; typically, this has a negligible effect on the light-collecting power of the telescope
(2) In a *Newtonian focus* arrangement, the image is reflected through a hole in the side of the telescope; this focal design is often employed in moderate-to-large sized personal telescopes (see *Focal Arrangements for Reflecting Telescopes*)
(3) In a *Cassegrain focus* arrangement, the image is reflected back down the telescope and out the bottom, through a hole in the center of the primary mirror; this arrangement is used in most large telescopes
(4) In a *Coudé focus* arrangement, the image is deflected, by means of three additional mirrors, to a fixed location remote from the telescope; this design is used when very large, immobile instruments are used to record and analyze light

D. Instruments for analyzing light sources
1. The most basic instrument is a *photometer*, which simply measures the brightness of a source of light
   a. A device called a photomultiplier commonly is used in astronomy
      (1) A *photomultiplier* is a vacuum tube containing a metallic surface that releases electrons when photons of light strike it, creating an electrical current
      (2) The current is proportional to the intensity of the incident light, so measuring the brightness of a source is a matter of measuring the current produced in the photomultiplier
   b. Filters help measure the brightness of an object in different bands of wavelengths, providing information about the color of the observed object
   c. A photometer records no information on the size or shape of a source of light
2. A camera records the image of an astronomical source, providing information on its size and shape
   a. The focal ratio of the camera, combined with the focal ratio of the telescope, defines the field of view and the angular scale of the images recorded (*focal ratio* refers to the ratio of the focal length to the diameter of the mirror that focuses the light)
   b. As in photometers, cameras are often equipped with filters so that images of objects in restricted and well-defined wavelength regions are obtained
   c. A camera must have a device for recording the image
      (1) Film was used historically to record images, and is still used in some applications
      (2) Most modern instruments use electronic detectors instead of film (see point 4. below)
3. A **spectrograph** spreads light out according to wavelength, forms a spectrum, and records it
   a. A prism may be used to disperse the light
   b. More often astronomers use a grooved surface called a *diffraction grating* to disperse the light
   c. The resolving power of a spectrograph is a measure of how much detail can be discerned in the spectrum
4. A *detector* is a device that records the intensity of light

Focal Arrangements for Reflecting Telescopes

The diagrams below show the arrangements of mirrors for the three primary types of focus.

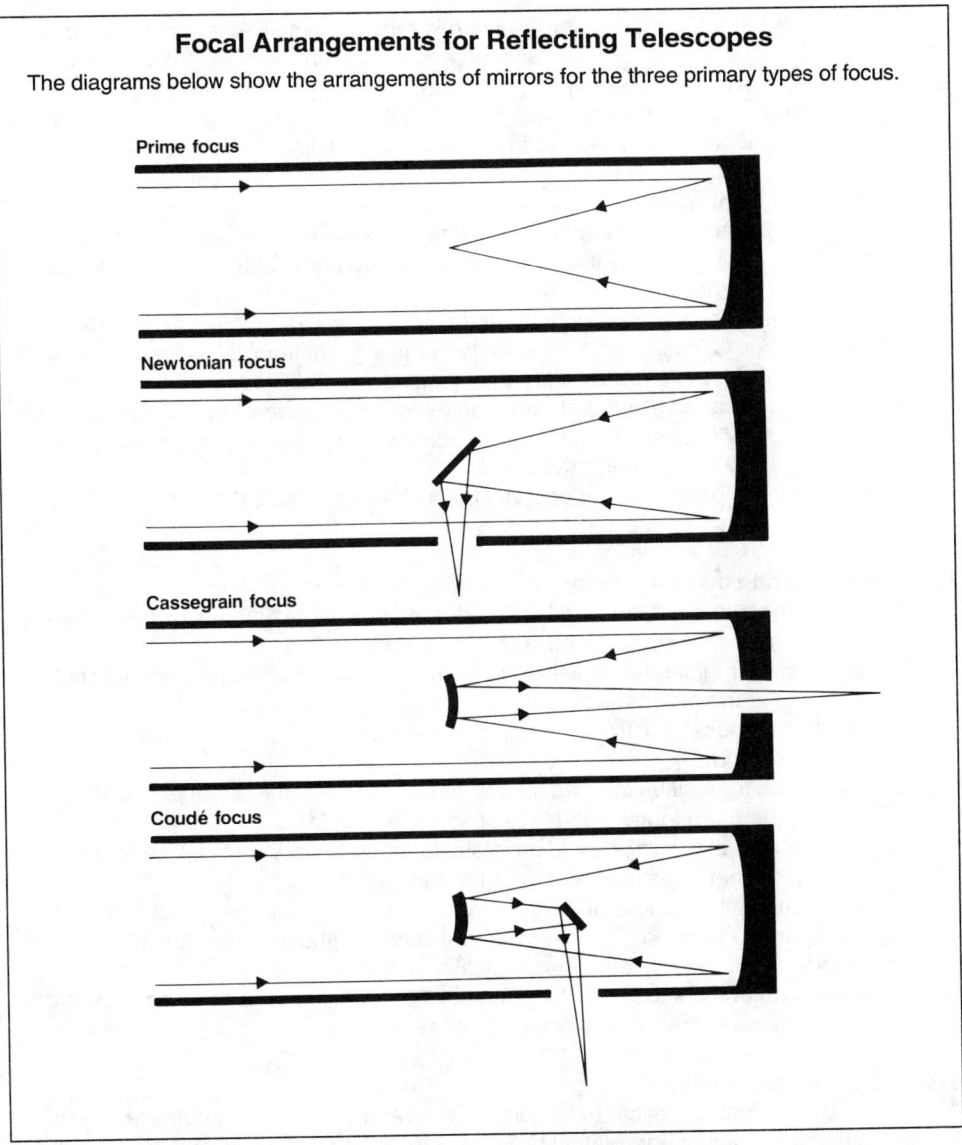

a. The individual picture elements or *pixels* in the detector limit the resolution of a camera or spectrograph
  (1) Pixel size, which is equal to the average size of the grains in the emulsion, typically is about 15 μm in film
  (2) For electronic detectors, the pixel size is normally larger than this; pixel sizes of 20 to 30 μm are common
b. Electronic detectors have many advantages over film
  (1) Detectors are far more sensitive, having up to 100 times the efficiency of film

(2) Detectors have a large *dynamic range*, normally being able to record images simultaneously and accurately that differ in intensity by up to a factor of one million
(3) Detectors are *linear*, meaning that the strength of the recorded signal is proportional to the intensity of incident light (whereas the response recorded on film is not uniformly proportional to the intensity of incident light)
(4) Detectors record intensities directly in the form of electrical signals, allowing observational data to be stored immediately in computers for analysis
c. The most widely used detector today is the *charge-coupled device (CCD)*
   (1) A CCD has a grid of fine wires creating electrical traps for electrons, which are freed by photons striking a metallic surface
   (2) The accumulated charges on the grid can be read out several times per second, as the charges are transferred row-by-row to the edge of the detector and recorded
   (3) The pattern of trapped electrical charges reveals the pattern of light intensities in the image

E. Observatory site considerations
1. A good observatory site has a large percentage of nights that are clear, is far removed from major cities, and is at high altitude
   a. Before a major telescope is built, years are spent gathering weather statistics at the proposed site
   b. The best locations from this point of view tend to be in arid, desert regions of the world
2. Air pollution from cities can reduce the transparency of the atmosphere, thus reducing the astronomer's ability to observe faint objects
3. A good observatory site should have stable air overhead (that is, little turbulence), so that atmospheric blurring is minimized
4. High altitude allows more of the light from the source to reach the telescope; this is especially important for near-infrared wavelengths, where many absorption bands are formed by atmospheric water vapor
5. The latitude of the site is an important consideration, because this determines which hemisphere of the sky can be observed

F. Major observatory sites
1. The U. S. National Optical Astronomy Observatories (NOAO), supported by the National Science Foundation (NSF), is the major facility available to U.S. astronomers, with sites at Kitt Peak, Arizona, and Cerro Tololo, Chile
2. Another major U.S. site is Mauna Kea, Hawaii, which houses several large telescopes—including the world's largest telescope, the 10-m Keck telescope—operated by various agencies
3. A consortium of nations operates the European Southern Observatory (ESO), whose instruments are located on La Silla, a mountain in the Chilean Andes
4. Other important large telescopes are scattered at various sites around the world

G. Future visible-light telescopes
1. The limiting factor in building any large telescope is the size of the primary mirror
   a. Traditionally, mirrors were cast in large molds

            (1) For very large mirrors it is difficult to have the glass cool without cracking or becoming distorted
            (2) Mirrors made in this manner are thick and heavy, creating engineering problems in supporting them rigidly and pointing the telescope accurately
            (3) The largest mirror ever made this way is a flawed Soviet 6-m instrument
        b. One alternative being used today is to build a large, thin mirror and then actively control its shape to counteract sag
            (1) Such a mirror has the advantage of being relatively light in weight, thereby minimizing the engineering difficulties for the telescope structure
            (2) Push-rods behind the mirror are used to flex it and keep its shape correct as the telescope is pointed in different directions
        c. Another new method for making large mirrors is to make them mostly hollow, with a honeycomb structure in the back
            (1) This minimizes weight while maximizing rigidity
            (2) These mirrors, being built at the University of Arizona, also use a rotating oven for casting, so that the glass takes on a concave shape and requires relatively little grinding to achieve its final shape
        d. Another strategy is to combine several small mirrors to obtain the light-collecting power of a single large one
            (1) This technique was used at the Multiple-Mirror Telescope on Mt. Hopkins, Arizona, which has six 1.8-m mirrors whose combined light-collecting power is equivalent to that of a single 4.5-m mirror
            (2) The 10-m Keck telescope has 36 individual hexagonal mirrors attached together, with each one activated to keep the entire ensemble focused as the telescope is pointed in different directions
    2. Several major instruments are in the planning stages or under construction, using these techniques
        a. A second 10-m Keck telescope is under construction on Mauna Kea, to be operated in conjunction with the first one
        b. The NOAO has started work on a pair of 8-m telescopes, one to be on Mauna Kea and the other near Cerro Tololo, in collaboration with several international partners
        c. The ESO is building a 16-m telescope, consisting of four 8-m instruments whose light will be combined

II. Telescopes for Invisible Wavelengths

A. General information
1. Visible light forms only a tiny fraction of the full electromagnetic spectrum; all astronomical observations were confined to this narrow spectral region up until about 50 years ago
2. Radio astronomy got its start in the 1930s, with small experimental receivers
    a. The first person to detect radio signals from astronomical sources was Karl Jansky
    b. With the exception of isolated work done individually (notably by Grote Reber), radio astronomy was not developed as a science until after World War II

3. Wavelengths other than visible and radio-wave radiation do not penetrate Earth's atmosphere and must be observed from space
4. Scientists launched the first space-based telescopes on small "sounding rockets" that provided only a few minutes of observations
   a. The first observations of ultraviolet radiation from the Sun took place in the late 1940s
   b. Pioneering ultraviolet and X-ray observations of stars took place using sounding rockets in the 1980s
   c. Sounding rockets still are used to develop new instrumental concepts and to carry out research that can be accomplished in short periods
5. Longer, space-based observations are accomplished from Earth's orbit
   a. The first astronomical satellites were launched in the early 1970s
   b. Currently, several space-based observatories are in operation, including the *Hubble Space telescope (HST)*

B. Radio telescopes
1. A radio telescope has a design similar to a reflecting telescope (used for visible light)
   a. A concave reflecting surface brings radio-wave radiation to a focus
      (1) This is the familiar radio antenna or satellite dish, such as those commonly seen at radar installations, communications centers, and in family back yards
      (2) In order to form precise images, the reflecting surface must be smooth relative to the wavelength being observed; because radio wavelengths are long, the reflector need not be very smooth and in some cases consists of a wire mesh
   b. At the focus of the antenna is a receiver that converts radio waves into electrical signals that can be recorded
2. It is possible to build very large radio telescopes because reflecting surfaces can be thin and lightweight; antennae have been built as large as 100 m for steerable dishes, and larger for fixed instruments, such as the 300-m radio telescope at Arecibo, Puerto Rico
3. Despite their large size, radio telescopes have poor resolution, because radio wavelengths are comparatively long
4. High-angular resolution can be achieved by combining signals from two or more radio telescopes in a technique called *interferometry*
   a. Interferometry relies on interference between waves received at separate antennae
      (1) Waves reach separate antenna at slightly different times; the time difference depends on the precise angle from which the radiation comes
      (2) The small time-of-arrival differences between waves received at separate antennae results in interference when the waves are combined electronically
      (3) The pattern and degree of interference can be used to infer the precise direction from which the waves are coming
   b. Angular resolutions of better than 0.1 arcsecond can be achieved with arrays of radio telescopes, such as the *Very Large Array (VLA)* in New Mexico
   c. The resolution that can be achieved depends on the separation of the antennae; thus a project called *Very Long Baseline Interferometry (VLBI)* is un-

der development, and will use radio telescopes distributed all over the world
5. Radio telescopes are useful for many kinds of research
   a. Ionized gas in space emits continuous radiation at radio wavelengths
   b. Hydrogen atoms in space emit at a wavelength of 21.1 cm, providing a very useful probe of the distribution of interstellar gas in our galaxy and in other galaxies
   c. Molecules in dense interstellar clouds form emission lines at short radio wavelengths (from a few mm to a few cm)
   d. Some stars are radio emitters because of ionized gas surrounding them or flowing out from them
   e. Many galaxies produce intense, continuous radio emission from their cores
6. Major radio observatories are located in several countries
   a. The U.S. National Radio Astronomy Observatory (NRAO) operates facilities in West Virginia (at Green Bank), in New Mexico (the *VLA*), and in Arizona (a 12-m dish for observing interstellar molecules, located on Kitt Peak)
   b. The giant Arecibo dish in Puerto Rico is operated by Cornell University, with support from the NSF
   c. Large radio telescope arrays operate in England, Holland, and Australia
   d. Smaller dishes, which are appropriate for the shorter wavelengths of molecular emissions, are operated in Arizona, Hawaii, Chile, and other countries
7. The *Cosmic Background Explorer (COBE)* is one radio observatory operating in space; this instrument has mapped the sky at wavelengths near 1 mm, where relic radiation from the earliest era of the universe is observed (see Chapter 17, Cosmology)

C. Ultraviolet telescopes
1. Because ultraviolet radiation cannot penetrate the Earth's atmosphere, all ultraviolet observatories must be launched above the atmosphere
2. The optical design of an ultraviolet telescope is basically the same as a visible-light telescope
   a. Most ultraviolet telescopes flown in space have been Cassegrain designs
   b. Mirrors that reflect visible light efficiently tend to absorb ultraviolet radiation, so special measures must be taken to retain ultraviolet reflectivity
      (1) Aluminum reflects well at ultraviolet wavelengths, but oxidizes very easily (even at the low densities in space), which destroys its reflectivity
      (2) Most ultraviolet telescopes use special transparent coatings designed to protect the aluminum from oxidation while preserving its reflectivity
3. Scientists have launched several ultraviolet missions, and more are planned
   a. The series of Orbiting Astronomical Observatories (OAO) included two very successful missions
      (1) OAO-2 operated between 1969 and 1972
      (2) OAO-3, also known as Copernicus, operated between 1972 and 1980
   b. The International Ultraviolet Explorer (IUE), a joint U.S.- U.K.-European Space Agency project, is still operating, some 16 years after its launch
   c. The HST has ultraviolet capabilities in its cameras and its spectrographs
   d. The Extreme Ultraviolet Explorer (EUVE) was launched in 1992 and has been mapping the sky at very short ultraviolet wavelengths not previously explored

e. The planned Far-Ultraviolet Spectroscopic Explorer (FUSE), expected to be launched in 2001, will provide coverage of ultraviolet wavelengths not accessible to the HST

D. Infrared telescopes
1. Some radiation in the infrared spectrum can penetrate the atmosphere; some infrared work is done using ground-based telescopes, but most detection takes place in space
2. The optical design of an infrared telescope is identical to a visible-light telescope, but there are some significant additional considerations
    a. The telescope, and especially the detector, must be kept cold in order to minimize thermal emission
        (1) The telescope and sky glow at the same wavelengths being observed, creating background radiation that can swamp the source being observed
        (2) Cooling the instrument helps minimize this interfering radiation
        (3) The cooling usually is provided by liquid nitrogen (77 K) or liquid helium (4 K) circulating through the instrument
    b. Elimination of sky background is normally accomplished by measuring the radiation from a point in the sky near the target, and subtracting this measurement from the measurement of the target, to obtain the actual radiation intensity of the target
3. The low energy of infrared photons (relative to visible or ultraviolet photons) makes them difficult to detect efficiently
    a. The lack of sensitive detectors until recently was the most important technological issue for infrared astronomy
    b. New electronic detectors now provide sensitivities for infrared observations comparable to those already available for other wavelength bands
4. Infrared observations are important in many types of astronomical research
    a. Relatively cool objects emit most strongly at infrared wavelengths, according to Wien's law (see Chapter 4, Light and the Atom)
    b. Newly formed stars and interstellar clouds in the process of collapsing (to form stars) are currently the subject of intensive infrared work
    c. Interstellar dust, consisting of tiny solid particles in space, glows at infrared wavelengths and can be mapped and studied using infrared telescopes
    d. The properties of gas flows from stars in many stages of their lives can be observed using infrared wavelengths
    e. The Infrared Astronomical Satellite (IRAS), an infrared telescope, discovered an entirely new class of galaxies, so shrouded in dust that they are seen only in infrared wavelengths; these starburst galaxies appear to be undergoing intense phases of star formation (see Chapter 16, Galaxies)
5. Major ground-based infrared telescopes have been built on Mauna Kea, where the extreme altitude reduces the amount of water vapor overhead
6. Some mid- to far-infrared observations are made from high-altitude balloon and aircraft observatories; the most important of these is the Kuiper Airborne Observatory (KAO), operated by the National Aeronautics and Space Administration
7. Scientists have launched one major space-based infrared observatory and others are planned

a. The IRAS, a joint Dutch-U.K.-U.S. observatory, operated in 1983 and 1984; it mapped the entire sky at four far-infrared wavelengths
b. The planned Infrared Space Observatory (ISO) will incorporate several different instruments, and will be launched by the European Space Agency (ESA) in 1997
c. The Space Infrared Telescope Facility (SIRTF), a U.S. observatory, is now planned for a post-2000 launch

E. X-ray telescopes
1. X-rays do not penetrate Earth's atmosphere and must be observed from space
2. The optical design used for visible, ultraviolet, and infrared telescopes does not work at X-ray wavelengths because mirrors do not reflect X-rays that strike at normal angles (that is, at angles nearly perpendicular to the mirror)
3. X-rays can be reflected if they strike the mirror surface at very small angles; this is called *grazing-incidence reflection*
   a. An X-ray mirror for a telescope consists of a thin ring designed so that X-rays are deflected at grazing angles to a focus
   b. Because such a ring has very little collecting area, an X-ray telescope normally consists of a nested array of concentric rings which have a common focal point
   c. An X-ray telescope can perform all the functions of a normal telescope; this is possible by constructing both grazing-incidence secondary mirrors and grazing-incidence diffraction gratings
4. Scientists can use X-ray observations in several important research areas
   a. According to Wien's law, an object that is hot (more than 100,000 K) will emit most strongly at X-ray wavelengths; thus, X-ray telescopes are especially useful in probing for high-energy activity in the universe
   b. X-ray telescopes can detect many kinds of stars with very hot outer regions known as **coronae**
   c. Much of the interstellar space in our galaxy is filled with rarefied, hot gases that emit at X-ray wavelengths
   d. Compact stellar remnants (such as white dwarfs, neutron stars, and black holes) create intense X-ray radiation; these objects are probed using X-ray telescopes
   e. Many clusters of galaxies are infused with very hot (100 million K) gas that glows at X-ray wavelengths
   f. Many galaxies and related objects, such as quasars, have intensely active cores that emit X-rays
5. Scientists have launched several X-ray observatories; others are planned
   a. The *Uhuru* satellite was the first to map the sky at X-ray wavelengths, and it discovered many strong sources
   b. Several U.S. and European X-ray observatories were launched in the 1980s and 1990s; some Japanese instruments also have been launched recently
   c. The next major X-ray observatory to be launched will be the X-Ray Timing Explorer (XTE), which will make accurate timing measurements of X-ray fluctuations from such sources as binary star systems and rotating neutron stars (see Chapter 14, Stellar Remnants, for more information)
   d. In the future, the Advanced X-Ray Astrophysics Facility (AXAF) is being developed for a 1998 launch

F. Gamma-ray telescopes
1. Gamma rays can be observed only from space
2. No mirror has been developed that will reflect gamma rays, so the design of a gamma-ray telescope is very different from that of any other kind of telescope
   a. Instead of bringing gamma rays to a focus, a gamma-ray telescope blocks out all incoming gamma rays except those coming from the direction in which the telescope is pointed
   b. While it is not possible to obtain gamma-ray images of objects, one can measure the location of the source, determining its intensity and spectrum
3. Gamma-ray observations have several important scientific goals
   a. Most gamma rays are emitted by nuclear processes
      (1) The protons and neutrons in an atom's nucleus have energy levels analogous to those of the electrons orbiting the nucleus
      (2) The energies are much higher, however, so released photons are extremely energetic and have wavelengths in the gamma-ray range
   b. Nuclear reactions in supernova explosions produce observable gamma-ray emission lines, which can be analyzed to determine the nature of the reactions
   c. Gamma-ray emission lines also are produced by spontaneous reactions in radioactive materials in space
   d. Gamma-ray emission lines can be produced by the impact of subatomic particles called **cosmic rays** on grains of interstellar dust
   e. Gamma rays also are produced when subatomic particles collide and annihilate each other; a particularly prominent line results from collisions between electrons and positrons in space (see Chapter 17, Cosmology)
   f. Scientists have identified a mysterious class of objects known as gamma-ray burst sources; these are not yet fully understood
4. Some gamma-ray observations are possible from high-altitude balloons; the great supernova of 1987 was probed using balloon-borne gamma-ray telescopes
5. Small French and American gamma-ray detectors now provide rough maps of the sky
6. The currently operating Gamma-Ray Observatory (GRO), also known as the Compton Observatory, has made many important observations of gamma-ray emission lines and has helped discover many gamma-ray emission sources

---

Study Activities

1. Explain the benefits of using telescopes rather than the unaided human eye for astronomical observations.
2. Summarize the designs of both refracting and reflecting telescopes; comment on their advantages and disadvantages.
3. Research the technologies being developed today for building very large mirrors for visible-light telescopes.
4. Explain the special precautions that must be taken in designing and operating an infrared telescope to defeat background radiation.
5. Compare and contrast the designs of X-ray and gamma-ray telescopes, and show how they differ from visible-light telescopes.

6

Solar System Formation and Planetary Science

Objectives

After studying this chapter, the reader should be able to:
- Describe the overall structure of the solar system.
- Summarize the general properties of the terrestrial and giant planets.
- Explain the modern theory for the formation of the solar system.
- Describe the internal structure and geological activity in terrestrial planets.
- Discuss the processes that govern atmospheric composition and circulation for both terrestrial and giant planets.

I. Overview of the Solar System

A. General information
1. The *solar system* consists of the Sun, nine known planets, the many satellites of the planets, and a large number of interplanetary bodies (such as comets, asteroids, meteoroids, and interplanetary dust particles)
2. Astronomers study the planets by telescopic observation from Earth; space probes have observed all but Pluto from close range
   a. The best images obtained from Earth can reveal some surface features on the planets, thereby providing information on geology and atmospheric circulation
   b. Spectroscopic observations from Earth provide information on atmospheric composition through the analysis of spectral lines formed in planetary atmospheres
   c. Radio and infrared measurements collect data on planetary surface temperatures (using Wien's law) and on the overall energy of the planets
   d. Space probes, primarily from the United States and the former Soviet Union, allow detailed studies of surface geology and atmospheric circulation
   e. Several current or planned missions to the planets will further increase our knowledge
      (1) *Galileo* is on its way to a late 1995 arrival at Jupiter
      (2) The *Cassini* mission will be launched in 1997 toward Saturn
      (3) A Pluto fly-by mission is now being planned for a launch at the end of this decade

B. Structure and motion of planets and moons
1. The solar system has an overall disk-like structure
   a. The orbital planes of the planets are all closely aligned with each other
   b. The orbital shapes are nearly circular
      (1) Another way to say this is that the orbits of most planets are ellipses with small *eccentricities*
      (2) The eccentricity of an ellipse is defined as the ratio of the separation between foci to the length of the semimajor axis
      (3) The small eccentricities of the planetary orbits represent sufficient departures from perfect circles that the ancient epicyclic model of motions was inaccurate
2. All planetary orbits, and virtually all satellite orbits, revolve in the same direction (counterclockwise as seen from above the north pole)
3. The direction of rotation for nearly all the planets and satellites is in the same direction as the orbital motions
   a. The rotation axis for most planets is nearly perpendicular to the orbital plane
   b. Venus, Uranus, and Pluto are exceptions, having their axes tilted more than 90° from a perpendicular to their orbital planes; if viewed from above the orbital plane, these planets would appear to rotate in the opposite direction from the other planets

C. Classes of planets
1. The **terrestrial planets** are the four innermost planets of the solar system, each one relatively small and dense with a rocky surface
   a. The average densities of the terrestrial planets range from roughly 3,500 to 5,500 kg/m^3 (3.5 to 5.5 g/cm^3)
   b. The composition of the terrestrial planets is relatively low in the lightweight, volatile gases, such as hydrogen and helium, and high in refractory elements, such as silicon and iron
   c. The terrestrial planets tend to have few or no satellites
2. The **giant planets**, or **gas giants**, are the outer planets (excluding Pluto), each of which is much larger than any of the terrestrials, has a low density, and no solid surface
   a. The average densities of the giant planets range between 700 and 1,700 kg/m^3 (0.7 and 1.7 g/cm^3)
   b. Lightweight, volatile gases, such as hydrogen and helium, dominate the giant planets; this also is the composition of the Sun and stars
   c. The giant planets have many satellites and *ring systems* (disks of small, icy particles or dust in their equatorial planes)

II. Modern Theory of Solar System Formation

A. General information
1. Historically, scientists have considered two general types of models for solar system formation
   a. In the *catastrophic theories,* it was believed that the solar system formed as the result of a singular cataclysmic event caused by external forces; in this view, planetary systems are rare

b. In the *evolutionary theories*, cosmologists believed that the formation of the solar system was the result of natural internal processes accompanying the formation of the Sun; in this view, planetary systems orbiting other stars would be common
 2. The model used today to explain the formation of the Sun and planets involves elements of catastrophic and evolutionary theories
   a. The overall process of planet formation is evolutionary, and planetary systems are thought to be common
   b. However, many of the details are thought to be the result of singular, catastrophic events

B. Beginnings in an interstellar cloud
 1. The solar system originated from a rotating, condensing cloud of interstellar gas and dust (see Chapter 15, The Milky Way)
   a. Interstellar space is permeated with a rarefied medium of gas, with tiny dust particles (similar to smog) mixed in
   b. In some regions, this material is concentrated into clouds
 2. Observations and theoretical studies show that under certain conditions an interstellar cloud may collapse under its own gravity
   a. This may result from random density fluctuations, which might create a localized region of sufficient density so that collapse occurs
   b. It may result from compression of the cloud, perhaps by the blast wave from an exploding star, thereby triggering the collapse
 3. As a cloud collapses, it flattens into a disk because of its rotation
   a. This flattened disk gives rise to the overall disk-like shape of the planetary system
   b. The disk that led to formation of the Sun and planets in our own system is referred to as the **solar nebula**

C. Formation of the Sun
 1. As the cloud collapsed, the infall of material was most rapid at the center
   a. As a result, a dense concentration of gas built up at the center while the outer portions of the cloud were still just beginning to fall in
   b. This dense concentration was the precursor to the Sun
 2. The central concentration heated up as a result of compression, and the heat (in the form of infrared radiation) escaped for some time
 3. After about 1 million years, the central concentration became dense enough to form molecular hydrogen
   a. Molecular hydrogen traps infrared radiation, so that the heat loss was slowed
   b. The central object, now a **protostar**, slowed its contraction to a gradual rate while it continued to heat
 4. Eventually, the protostar became too hot for molecular hydrogen to survive
   a. The molecules were destroyed, and infrared radiation once again could escape
   b. This allowed the protostar to undergo a new rapid collapse phase
 5. The collapse was slowed again as gas pressure inside the protostar became sufficient to nearly balance the inward force of gravity; the protostar entered a new phase of very slow contraction
 6. After about 100 million years, nuclear reaction ignited in the core of the protostar, and the Sun was born

D. Formation of the planets and interplanetary bodies
1. Material from the solar nebula was still slowly falling in after the Sun was formed
2. Gradually, the gas in the nebula began to condense into solid form as it cooled
    a. The temperature was highest in the inner regions of the nebula, which were immersed in the powerful radiation field of the young Sun
        (1) The high temperature in the inner region prevented lightweight, volatile gases, such as hydrogen and helium, from condensing into solid form
        (2) Heavier, refractory elements (such as metals and rocky materials) condensed despite the high temperatures of the inner nebula
    b. In the outer regions where it was much colder, all elements (including the light ones) were able to condense
3. Soon the tiny solid grains that first formed began to accumulate into larger bodies
    a. These objects, ranging in diameter up to several hundred kilometers, are called **planetesimals**
    b. Eventually, the solar nebula consisted of a swarm of these bodies orbiting the Sun in a disk
4. The planetesimals then began to collide with each other, sometimes sticking together, thereby gradually forming the planets
5. Contrasting conditions in the inner and outer portions of the solar system led to the two contrasting classes of planets
    a. The planets formed in the inner part of the disk were small and dense, because they lacked the abundant lightweight elements
    b. In the outer part of the solar system, the newly formed planets had high masses and low densities
        (1) In the cold outer nebula all the elements (including the abundant lightweight ones) were incorporated into planets
        (2) Because of their high masses, the outer planets were able gravitationally to attract even more matter from the remaining gas and dust of the solar nebula
    c. The giant outer planets attracted nebular material that formed disks around them, giving rise to many moons and rings; this did not occur in the inner part of the solar system
6. Details and irregularities of some of the planets were the result of random events
    a. Late collisions between planetesimals are thought to have caused the unusual tilts of Venus, Uranus, and possibly Pluto
    b. The formation of our Moon is thought to be the result of a collision between Earth and a very large planetesimal
    c. Mercury may have lost much of its outer portion because of a similar collision
    d. All planets were heavily bombarded with leftover debris in the young solar system; many of the resulting craters are still visible today
7. Some of the planetesimals were not incorporated into planets
    a. In the region between Mars and Jupiter, a large number were prevented from merging together because of tidal forces exerted by the young Jupiter; these objects are known as **asteroids** or **minor planets**
    b. In the outer solar system, some of the original, icy planetesimals were never incorporated into planetary bodies, but remain today as giant satellites or free, Sun-orbiting bodies
    c. Pluto is thought to be such a body, consisting of roughly half ice and half rocky material

d. Many of the large satellites of Jupiter, Saturn, and Neptune are probably similar bodies that were captured into orbit about their parent planets
e. Scientists believe that many more minor planetary bodies are orbiting beyond Uranus, extending the disk of the solar system to some 50 astronomical units (AU) and serving as the source for some comets; this outer disk is called the **Kuiper belt**
f. Much farther out (roughly 100,000 AU) there exists another cloud of leftover debris from the formation of the solar system; this spherical cloud, called the **Oort cloud,** is the source of more comets

III. Planetary Science

A. General information
1. Several general processes and principles apply to all planets
2. All planets began from the same material, but various processes have defined and altered their individual characteristics

B. Processes that govern the planets' internal structure
1. Each planet is in balance between the inward force of gravity and internal forces that act outward
   a. This balance is called **hydrostatic equilibrium**
   b. In the terrestrial planets, the outward force is supplied by the rigidity of the rocky materials of which the planet is made; fluid pressure provides this force in cases where there are internal liquid layers
   c. In the gas giants, fluid pressure provides the outward force that counterbalances gravity
2. A process called **differentiation** causes relatively heavy elements to sink toward the center of each planet
   a. This process requires a fluid medium, implying that even the terrestrial planets have been at least partially fluid (molten) at some time in their past
   b. Differentiation in the terrestrial planets has helped to create a layered structure consisting of a thin *outer crust,* an intermediate-density (semi-rigid) *mantle*, and a dense nickel-iron *core* (which is partly fluid in some cases)
   c. In the gas giants, differentiation has created a small, solid core beneath the fluid layers that make up most of the interior
3. Internal circulation, driven by planetary rotation, occurs in any planet that has fluid zones in its interior
   a. In the terrestrial planets, the circulation occurs only in the core, if at all
   b. The interiors of the gas giants undergo complex circulation
4. Magnetic fields and particle belts are manifestations of internal circulation
   a. Electrical currents create a magnetic field; internal fluid motions in a planet can create electrical currents
      (1) Fluid motions in Earth's nickel-iron core have created a significant magnetic field
      (2) Mercury has a weak magnetic field despite the planet's slow rotation; astronomers therefore believe that this small planet has a large, partially molten core
      (3) All of the gas giants have intense magnetic fields due to internal circulation

b. A planetary magnetic field can trap electrically charged particles in belts or zones surrounding the planet
   (1) This is the origin of Earth's charged-particle zones, called the **Van Allen belts**
   (2) The giant planets have stronger magnetic fields and very dense particle belts

C. Surface features of the terrestrial planets
1. Slow-flowing motions in the mantle are responsible for *tectonic activity*, which is the movement of crustal plates and is associated with such phenomena as earthquakes and volcanism
   a. These flows in the mantle may be caused by **convection,** the overturning motion caused by heated bubbles rising in a gravitationally confined fluid
   b. On Earth, tectonic activity causes continental drift, while on Venus the crustal plates are locked into place and do not drift; little is known about tectonic activity on Mercury or Mars
   c. Tectonic activity is responsible for the major structures of the crusts of the terrestrial planets, such as mountains and continental distributions
2. Several conditions influence surface geology
   a. The composition of the crust is determined by the original composition of the planet (resulting from differentiation) and by volcanism, which supplies material from the interior
   b. Erosion by groundwater modifies surface structures formed by tectonic activity
   c. The surfaces of all terrestrial planets have many craters as a result of collisions with other bodies in space

D. Composition and circulation of planetary atmospheres
1. A balance between gains and losses of various constituents controls planetary composition
   a. Gains include the accretion of gases during and after planetary formation, the venting of gases from the interior, and chemical processes (including biological ones) that occur on the planet's surface
   b. Losses include the escape of gases into space (which is determined by the weight of the molecules and the temperature of the atmosphere) and chemical or biological reactions at the surface
2. Similar factors in all planets control the circulation of an atmosphere
   a. Convection, the tendency for warm gas to rise, creates high- and low-pressure zones and vertical flowing motions
   b. Planetary rotation converts simple convective flows into rotary flow systems ranging from gentle weather patterns to violent storms on the terrestrial planets and high-speed, planet-circling flows on the gas giants

Study Activities
1. Summarize the kinds of information that can be obtained by making Earth-based telescopic observations of planets.
2. List the aspects of planetary motion and properties that do not fit into the general pattern set by the majority of the planets.

3. Summarize the contrasts between terrestrial and giant planets.
4. Explain the difference between a catastrophic theory and an evolutionary theory of solar system formation. Discuss how the modern, accepted theory fits or does not fit into each category.
5. Summarize the process of formation of the Sun.
6. Describe how the modern theory of solar system formation accounts for the disk-like overall structure of the solar system; the fact that all the planets orbit the Sun in the same direction and approximately in the same plane; the fact that the outer planets tend to have many satellites while the inner ones have few or none; and the unusual tilts or spins of Venus, Uranus, and Pluto.
7. Discuss the role of differentiation in governing the internal structure of terrestrial and giant planets.
8. Summarize the various aspects of tectonic activity that shape the surface of a terrestrial planet.
9. List the factors that govern the composition of a planetary atmosphere.
10. Explain the factors that control the circulation of planetary atmosphere.

7

Earth and the Moon

Objectives

After studying this chapter, the reader should be able to:
- Explain the external appearance of Earth as seen from space.
- Describe the composition and vertical structure of Earth's atmosphere.
- Explain the circulation of Earth's atmosphere.
- Describe the interior structure of Earth and how it is probed.
- Describe Earth's magnetic field and its charged particle belts.
- Discuss the processes that shape Earth's crust.
- Describe the surface features of the Moon.
- Discuss the interior structure of the Moon.
- Summarize the formation and geological history of Earth and the Moon.

I. Earth

A. General information
1. Earth is the largest and most geologically active of the terrestrial planets; its oceans and nitrogen-oxygen atmosphere are unique among all planets
2. The large size of the Moon relative to Earth leads some to refer to the Earth-Moon system as a double planet

B. Earth's atmosphere
1. Earth's current atmosphere is considered a secondary atmosphere because it is not the original one
   a. The early atmosphere consisted primarily of hydrogen-bearing compounds and lacked oxygen
   b. Today's atmosphere consists of nitrogen (77%), oxygen (21%), and trace amounts of water vapor, argon, and other gases
   c. Life-forms on Earth are responsible for the composition of today's atmosphere
      (1) Plants consume carbon dioxide and release oxygen
      (2) The decay of organic matter (along with volcanic venting) produces nitrogen
2. The atmosphere can be divided into vertical layers, based on heat gains and losses
   a. Sunlight is the primary source of heat for the atmosphere, but it is absorbed by different components of the atmosphere at different layers

b. In the lowest layer of the atmosphere, called the *troposphere*, heating occurs at the bottom because of the absorption of sunlight at the surface, and temperature decreases with altitude
c. Above the troposphere is the *stratosphere*, a layer between 10 and 50 km in altitude where the temperature increases with altitude as a result of absorption of sunlight by ozone in the atmosphere
d. Above the stratosphere lies the *mesosphere*, between 50 and 80 km in altitude, where the ozone concentration is highest at the bottom and the temperature again decreases with altitude
e. The *thermosphere*, above 80 km in altitude, is a very rarefied region where the temperature gradually rises with altitude; here the heating at the upper levels comes from the impact of charged particles from space
3. The atmosphere undergoes complex flowing motions, creating climate and weather patterns in the troposphere
   a. Convection causes air to rise in warm regions, flow toward cooler areas, and descend
   b. Low pressure regions are formed in areas where air is rising, while high pressure forms where air is descending
   c. The temperature contrasts between continents and oceans, which change with the seasons, complicate the convection pattern
   d. Earth's rotation causes flowing air to be diverted into rotary patterns
      (1) The direction of flow depends on the local pressure variations and on the hemisphere
      (2) In the northern hemisphere, air circulates counterclockwise around a low-pressure region and clockwise around a high-pressure region; in the southern hemisphere, air circulates in the opposite direction

C. Earth's interior
1. The internal structure of Earth is probed using **seismic waves,** vibrations caused by earthquakes, which travel through the interior
   a. Waves involving vibrations perpendicular to the direction of travel are called *transverse waves* or *S waves* by geologists; they can pass through the solid earth but not liquid zones
   b. Waves involving vibrations along the direction of travel are called *compressional waves* or *P waves* by geologists; they can travel through both solid and liquid zones
   c. A third type of wave, also a transverse wave, is called *L wave* by geologists; these waves travel over Earth's surface only
   d. By mapping and timing the arrival of the different types of seismic waves at locations all over Earth, geologists can infer the locations and densities of internal solid and liquid zones
2. Earth's interior is divided into several distinct zones
   a. The outermost zone, which varies in depth from 5 to 60 km, is the *crust;* it consists of rock and surface soils
   b. Below the crust is the *lithosphere*, the uppermost portion of the *mantle;* this is a semi-rigid zone extending to a depth of nearly 3,000 km
   c. The central portion of the mantle is the *asthenosphere;* in this zone, slow flowing motions occur
   d. The lower portion of the mantle is the *mesosphere;* this zone is more rigid
   e. The *core* of Earth consists of a liquid outer zone and a solid inner region

3. Earth's magnetic field is generated in the outer core
   a. Motions of molten material in the outer core create electrical currents that give rise to Earth's magnetic field
   b. The field is *bipolar*, meaning that it has only two poles, connected by lines of magnetic force
   c. The magnetic force lines extend out into space, where they trap charged particles in zones called the **Van Allen belts**
   d. Some of these charged particles enter the upper atmosphere where the magnetic force lines approach Earth's poles; these particles create the aurorae
   e. Geological evidence indicates that Earth's magnetic field reverses direction on irregular timetables of tens to hundreds of thousands of years
      (1) The best evidence for this comes from embedded magnetic orientations in rocks from the seafloor formed at different times
      (2) These reversals are thought to be caused by irregularities (instabilities) in the fluid motions in Earth's outer core

D. Rocks
1. Rocks can be classified according to their origin and form
   a. *Igneous rocks* are formed from molten material resulting from volcanic eruptions
   b. *Sedimentary rocks* are formed by the deposition and hardening of layers of silt and debris in lakes and oceans
   c. *Metamorphic rocks* are altered and shaped by heat and pressure beneath the surface
   d. Rocks can be modified and converted from one type to another in various geological processes
2. Rocks also can be classified according to their mineral content
   a. About 90% of all rocks are silicates, with a composition dominated by silicon-oxygen compounds; common examples are basalts, granites, and quartzes
   b. Oxides, carbonates, and sulfides also are common
   c. The structure of most minerals consists of *crystals*, which are regular arrangements of the atoms in three-dimensional lattices

E. Changes in Earth's crust
1. The crust consists of thin, brittle plates floating on top of the denser mantle; the continental crust is the least dense and therefore floats the highest
   a. About a dozen major tectonic plates are active; the results of their motion are called **tectonic activity**
   b. Currents in the mantle (the asthenosphere) cause the continents to move around, changing their configuration on timetables of hundreds of millions of years
   c. At plate boundaries, mountain ranges rise as a result of compression
      (1) Deep undersea trenches form as a seafloor plate slides under a continental plate in a process called *subduction*
      (2) Earthquakes and volcanoes occur as the plates slide under each other
2. Volcanic eruptions create major changes in surface structures
   a. Many major mountains throughout the world near tectonic plate boundaries are volcanic in origin
   b. Extensive regions in many locations all over Earth are built up from lava flows

c. Many surface rocks are products of volcanic eruptions or have been modified by them
3. Erosion causes changes in surface features
   a. Glaciers have advanced and retreated over large portions of Earth, grinding away mountains and smoothing the terrain
   b. Weather, especially rainfall, flowing water, and wind, has eroded and smoothed many surface features, creating valleys and canyons
   c. Collisions also have modified surface features
      (1) Earth was bombarded heavily early in its history, along with the other planets
      (2) Most craters have been obliterated by erosion, tectonic activity, or volcanism, but some remain
      (3) Impacts still occur occasionally, posing a threat of major disruption to society and the climate

II. The Moon

A. General information
1. The Moon is very large relative to its parent planet, Earth; its diameter is a little larger than one-quarter of Earth's diameter, and its mass is over 1% of Earth's
2. The Moon has significant effects on Earth: its tidal forces cause the tides, and its appearance and cycles of phases and eclipses have had profound cultural influences
3. The Moon always keeps the same side facing Earth
   a. The rotation period of the Moon exactly equals its orbital period; this is called **synchronous rotation** (see Chapter 3, The Evolution of Modern Astronomy)
   b. Earth's tidal forces have slowed the Moon's rotation until the rotation and orbital periods became equal
4. Therefore, the far side of the Moon was not explored until the first unmanned missions photographed it in the 1960s
5. Telescopic observations as well as unmanned and manned exploration have yielded detailed information about the Moon's structure and origin

B. The lunar surface
1. The most prominent features on the Moon are the large, dark **maria** (once thought to be seas); the extensive, brighter highland regions; and innumerable craters of all sizes
2. The maria are concentrated on the side of the Moon that always faces Earth
   a. The lunar crust is thinner and denser on the near side than on the far side
   b. Therefore, Earth's tidal forces acted to "lock in" the Moon's synchronous rotation with the heavy side facing Earth
3. The maria are covered by extensive, ancient lava flows
   a. These low-lying areas have fewer craters than the rest of the surface, thus indicating that they are younger
   b. Radioactive dating of rocks from the maria indicate that they are around 3 billion years old
   c. The lava flows that resulted in the maria may have been triggered by large collisions at a time when the lunar interior was still partly molten

d. Meandering solidified lava rivers called *rilles* are seen occasionally in the maria
4. The lunar highlands are very ancient regions, densely covered with craters, and lacking erosion and drainage features; they are 3.5 to 4.5 billion years old, almost as old as the Moon itself
5. Collisions, not volcanoes, formed the lunar craters
   a. The craters have features, such as elevated rim walls and central peaks, that are characteristic of impact craters
   b. In many cases, *rays* are seen radiating away from a crater; these are streaks of material ejected by the impact of the collision
   c. Lunar craters range in size from *micrometeorite craters* a few millionths of a meter in diameter to enormous basins several hundred kilometers across
   d. Most craters were formed during the first billion years of lunar history, when the solar system was still permeated by leftover debris from its formation; the lunar craters remain because there is no atmospheric erosion
6. The lunar surface material consists of soil and loose rock fragments
   a. The soil, called the *regolith,* consists of rock fragments and small mineral deposits and extends to a depth of roughly 10 m
   b. The rocks on the lunar surface primarily are *breccias,* irregular chunks of material resembling pieces of concrete; these consist of small rocks fused together by the heat of meteorite impacts

C. The interior of the Moon
1. Collisions resulting from discarded spacecraft crashing onto the Moon have created seismic waves, allowing scientists to probe the Moon's interior
2. The lunar interior consists of a crust that is 50 to 100 km thick; a mantle consisting of a rigid lithosphere and a semi-rigid asthenosphere; and a core that is probably solid
3. No detectable lunar magnetic field exists; this is additional evidence against a liquid zone in the core
4. Tectonic activity is not evident on the Moon; it appears to be geologically inert

III. Development of the Earth-Moon System

A. General information
1. The Earth-Moon system is unique in the solar system
   a. With the exception of Pluto and its satellite Charon, the Earth-Moon system has the smallest contrast between the size and mass of a parent planet and its moon
   b. Earth is the only terrestrial planet to have a substantial satellite of any kind; the two tiny moons of Mars are thought to be captured asteroids
2. The large mass of the Moon, along with its large orbital separation from Earth, means that the Moon has very high angular momentum
   a. Angular momentum is the product of the mass, orbital radius, and orbital speed in the case of a circular orbit (for a more detailed definition, see Chapter 3, The Evolution of Modern Astronomy)
   b. Thus, the Moon's high angular momentum means that it represents an unusually large mass orbiting at an unusually large distance from its parent planet

  c. The high angular momentum of the Moon is not explained easily by many early theories of the Moon's origin
 3. Examination of lunar soil samples showed some significant chemical contrasts with Earth rocks, which any successful theory of lunar origins must explain
  a. The Moon has a low amount of iron and related elements, compared to Earth
  b. Volatile elements (those most easily vaporized) also are scarcer on the Moon
  c. Certain atomic forms or isotopes of elements, such as oxygen, are present on the Moon in the same relative abundances as on Earth
 4. Historically, three contrasting theories of lunar formation have been considered
  a. In the *fission theory*, the Moon was thought to have split off from Earth at an early time
   (1) This theory is ruled out because it does not explain the high angular momentum of the Moon; that is, the theory cannot explain how such a large mass could have been ejected into such a large orbit
   (2) It also does not explain the chemical contrasts between Earth and the Moon
  b. In the *capture theory*, the Moon was formed elsewhere and then captured by Earth's gravity
   (1) Such a capture is very difficult to achieve and hence very unlikely to have occurred
   (2) This theory does not explain the similarities in oxygen isotopes between Earth and the Moon
  c. In the *coeval formation theory*, Earth and the Moon were formed at the same time
   (1) This theory does not explain the high angular momentum of the Moon
   (2) It also does not explain the chemical contrasts between Earth and the Moon

B. The origin of Earth and the Moon
 1. Earth and the Moon, along with the other planets, are thought to have formed from the coalescence of planetesimals; these were intermediate-sized solid bodies that condensed from the gas and dust in the early solar system
 2. This modern hypothesis states that after Earth had formed and differentiated, it underwent a grazing collision with a large planetesimal (about the size of Mars)
  a. The collision ejected a large amount of material from Earth's outer layers, which were iron-deficient because Earth had already differentiated
  b. The ejected material was heated by the impact of the collision, thereby vaporizing much of the volatile material present
  c. The ejected material formed a disk around Earth that condensed to form the Moon
  d. The high angular momentum of the Moon is due to the high speed of the impacting planetesimal
  e. This theory fits all the observed facts; it is basically a blend of the earlier capture and fission theories

C. Subsequent evolution of the Moon
 1. The Moon was largely molten initially, after it formed from the debris orbiting Earth
 2. Soon after the lunar crust had cooled and hardened, large collisions broke through it, thereby triggering the vast lava flows that created the maria in areas where the crust was relatively thin

      a. Thus, the maria are younger (by about a billion years) than the lunar highland regions

      b. The lava flows obliterated existing craters, so the maria have fewer craters than the highlands

3. The Moon's interior cooled and solidified, and there has been little geological activity on or in the Moon since then
4. Tidal forces exerted by Earth on the Moon have slowed the Moon's rotation

      a. As the Moon's spin has slowed, the Moon has moved gradually farther away from Earth, maintaining constant angular momentum

      b. The Moon eventually became "locked in," with the denser side (the side with most of the maria) permanently facing Earth in synchronous rotation

Study Activities

1. Explain why Earth's atmosphere is divided into vertical layers based on the way the temperature varies with altitude.
2. Explain how convection and Earth's rotation create the flow patterns of the atmosphere.
3. Explain why Earth's atmosphere today is not the same as Earth's original atmosphere.
4. Describe the three types of seismic waves and how they are used to determine the structure of Earth's interior.
5. Summarize the nature of tectonic activity, including an explanation of continental drift, earthquakes, and volcanism.
6. Explain why the craters on the Moon resulted from collisions instead of volcanic eruptions.
7. Explain why few maria are found on the far side of the Moon.
8. Describe the modern theory of the formation of the Moon, including the observable facts that support it.

8

Mercury, Venus, and Mars

Objectives

After studying this chapter, the reader should be able to:
- Summarize the general properties of Mercury, Venus, and Mars.
- Explain the unusual rotation and orbit of Mercury.
- Describe the contrasts between Mercury and the Moon, which bear a strong resemblance to each other.
- Explain why the atmosphere of Venus is so hot and dense.
- Compare and contrast the geology of Venus with that of Earth.
- Explain differences and similarities between Earth and Venus.
- Describe the atmosphere and the seasons on Mars.
- Summarize the evidence that Mars once had plentiful water on its surface.
- Summarize the geological history of Mars.
- Describe the search for life-forms that was conducted on Mars by the *Viking* missions.

I. Mercury

A. General information
1. Mercury is the innermost of the nine planets
2. Mercury is very small compared to Earth (its radius is a little more than one-third that of Earth) and bears a strong physical resemblance to the Moon
3. As seen from Earth, Mercury is never more than 28° away from the Sun and therefore can be observed only when it rises just before the Sun rises or sets just after the Sun sets
4. Most of what we know about Mercury results from the *Mariner 10* probe, which flew past the planet at close range three times in the mid 1970s

B. The spin and orbit of Mercury
1. Mercury's orbit is more elongated (that is, eccentric) than the ellipses of the other planetary orbits; at its greatest distance from the Sun, Mercury is more than 50% farther from it than when it is closest to the Sun
2. Mercury's rotational period is equal to two-thirds of its orbital period
    a. Therefore, every time Mercury passes through its point of closest approach to the Sun, the same side of Mercury, or precisely the opposite side, faces the Sun
    b. The Sun's tidal force on Mercury has slowed the planet's original rotation so that this alignment occurs at every close passage

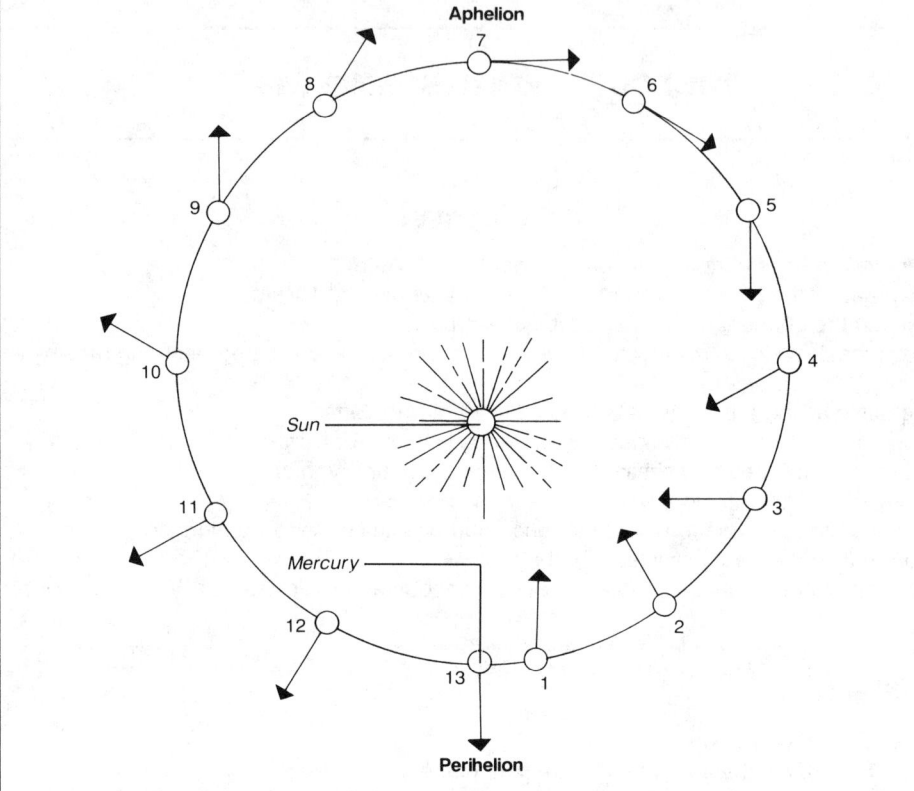

Spin-Orbit Coupling

Mercury rotates one and a half times as it completes one orbit around the Sun. Tidal forces helped "lock" Mercury into a 3:2 coupling; Mercury's heavy side is balanced when facing either directly toward or directly away from the Sun at its closest approach.

   c. This phenomenon is known as **spin-orbit coupling**, with a ratio of three spins for every two orbits (3:2 coupling; see *Spin-Orbit Coupling*); the simplest case is synchronous rotation, where the spin and orbital periods are equal (1:1 coupling)
   d. The reason Mercury is locked into a 3:2 coupling is that the tidal force caused by the Sun is much stronger at the point of closest approach than at any other point in Mercury's eccentric orbit, and Mercury itself has a heavy side that is balanced only when facing either directly toward or directly away from the Sun at its closest approach

C. The surface of Mercury
   1. Like the Moon, Mercury has a barren surface covered with lava flows, highland regions, and craters
   2. Significant contrasts also exist between the surfaces of Mercury and the Moon
      a. Mercury has extensive cliff or *scarp* systems, consisting of very long breaks in the crust
         (1) The Moon does not have these features

             (2) The scarps are thought to result from venting and shrinkage of the interior of Mercury after the crust hardened, thereby causing the crust to settle and crack
        b. Craters on Mercury and those on the Moon also differ
             (1) Craters on Mercury have lower rim walls than lunar craters
             (2) Ejecta from craters do not travel as far from the point of impact
             (3) These characteristics are caused by Mercury's stronger surface gravity
    3. An enormous impact basin called Caloris Planitia ("Plain of Heat") can be found on the side of Mercury that faces the Sun at closest approach every other orbit
        a. This is the hottest point in the solar system on alternate orbits of the Sun
        b. The embedded object that struck Mercury at this location may have created the asymmetric mass distribution causing the planet to lock into its 3:2 spin-orbit coupling

D. The interior of Mercury
    1. Relatively little is known about Mercury's interior because astronomers cannot use seismic waves to probe it
    2. From the high average density of the planet, astronomers infer that the core is very large, extending nearly three-fourths of the way from the center to the surface
    3. Despite its slow rotation, Mercury has a detectable magnetic field; this lends further support to the hypothesis of a large, partially molten core
    4. The lack of an extensive mantle suggests that Mercury may have undergone a collision with a very large planetesimal shortly after it formed, stripping away much of the original outer layer of the planet

II. Venus

A. General information
    1. Venus is the second planet from the Sun, orbiting at about 72% of Earth's distance
    2. Venus is never farther than 47° east or west of the Sun, so it is seen only in the evening or morning
    3. Venus is entirely covered with clouds, making observations of its surface from afar impossible
    4. Venus is similar to Earth in overall characteristics, such as mass and diameter, but the surface conditions on Venus are very different
        a. The temperature (730 K) is hot enough to melt lead
        b. The atmospheric pressure is nearly 90 times that at sea level on Earth
    5. Venus rotates very slowly (once every 243 days) in the retrograde direction (that is, backwards relative to the other orbital and spin motions in the solar system); this may be the result of a collision with a large planetesimal early in its history
    6. Venus has been probed by several unmanned spacecraft from the United States and the former Soviet Union
        a. The United States *Mariner* and the Soviet *Venera* missions obtained close-up photos and probed the atmosphere; *Venera* succeeded in landing craft on the surface, making measurements and obtaining photos
        b. The U.S. *Pioneer-Venus* mission orbited the planet for 14 years (1978 to 1992), observing variations in its atmosphere

c. The U.S. *Magellan* mission (1990 to 1992) recently completed a 2-year orbital study of Venus in which radar was used to obtain a detailed map of the entire surface

B. The atmosphere of Venus
  1. The atmosphere is composed primarily of carbon dioxide (CO_2), with small amounts of nitrogen, argon, and trace gases (including sulfuric acid [H_2SO_4]), which forms the clouds and precipitates out in the form of rain
  2. This atmosphere creates the extremely high surface temperature through the **greenhouse effect**
      a. The atmosphere allows sunlight to penetrate, thereby heating the surface
      b. The surface then emits infrared radiation
      c. The atmospheric gases trap infrared radiation, thus keeping heat bottled up in the lower atmosphere
  3. High-speed winds in the upper atmosphere circle the entire planet in about 4 days; the winds move in the same direction as the planet's rotation
  4. The clouds form three distinct layers at separate altitudes (48, 53, and 60 km above the surface), where the combination of pressure and temperature allows sulfuric acid to condense into droplets
  5. Variable levels of sulfur dioxide (SO_2) in the atmosphere observed during the *Pioneer-Venus* mission indicate that volcanic eruptions may be currently occurring on Venus

C. The surface and interior of Venus
  1. Rolling plains cover about 65% of the surface; about 25% of the surface consists of lowlands, and the highland regions make up the remaining 10%
  2. Impact craters are common on Venus
      a. They are distinguished from volcanic craters by their shapes, central peaks, and rim walls
      b. Only large craters exist on Venus because small meteoroids cannot penetrate the thick atmosphere and reach the ground
  3. The entire surface shows indications of volcanic activity, in the form of mountains with summit craters, lava flows, and lava streams, among other structures
      a. From the highly detailed radar maps it is not possible to confirm current volcanic activity, but some of the structures appear to be relatively young
      b. Circular structures called *coronae,* unique to Venus, are probably caused by upwelling lava
      c. In other lava flow regions, cracked patterns due to volcanic uplift are seen; these are called *tesserae*
  4. Tectonic activity on Venus has involved less movement of crustal plates than on Earth
      a. Because the crust of Venus appears to be more rigid than Earth's, crustal plates are not as free to move
      b. Many surface features appear to have resulted from tectonic activity; these include uplifted mountains, linear ridge-and-trench systems, and horizontal grooves formed by lateral stress or compression
  5. Because the diameter, mass, and average density of Venus are similar to those of Earth, astronomers expect that the internal structure is similar
      a. No seismic probes of Venus have been possible, so the internal structure has not been determined directly

b. Venus lacks a detectable magnetic field, but this may be due to its slow rotation rather than the absence of a partially molten core

D. The evolution of Venus
1. The widely differing surface conditions between Venus and Earth can be attributed to their differing distances from the Sun
2. Whereas Earth had liquid oceans from very early times, the increased intensity of sunlight on Venus was sufficient to prevent water from remaining in the liquid state
    a. Some scientists think that Venus might have had liquid oceans initially, as a result of accretion of cometary bodies from space
    b. Even if there were early oceans on Venus, the intense sunlight would have evaporated them, releasing the water to the atmosphere in the form of water vapor
3. Water vapor creates a mild greenhouse effect, which further increases the temperature on Venus
4. The oceans on Earth removed CO_2 from the atmosphere by dissolving CO_2 and depositing it in carbonate rocks; this did not happen on Venus because there were no oceans
5. CO_2 therefore accumulated in the atmosphere of Venus, elevating the temperature through the greenhouse effect
6. Eventually, the atmospheric temperature became so hot that water vapor dissociated, and the resulting free hydrogen atoms escaped into space, leaving Venus with the dense carbon dioxide atmosphere it has today
7. Some scientists are concerned that the runaway greenhouse effect that made Venus so inhospitable could occur on Earth if enough CO_2 is released into Earth's atmosphere

III. Mars

A. General information
1. Mars is the fourth planet from the Sun, orbiting at about 1.5 times Earth's distance
2. Mars is considerably smaller than Earth and Venus and has only a thin atmosphere
3. It has two tiny moons, Phobos and Deimos, which were discovered in 1877
4. Surface markings on Mars led some early observers to hypothesize the existence of a global system of canals
    a. Thus Mars has been the subject of intense speculation about possible civilizations or alien life-forms
    b. At the turn of the century, American amateur astronomer Percival Lowell wrote extensively on the details of the purported Martian civilization
    c. The continuing fascination with life on Mars was illustrated by the famous Orson Welles radio broadcast in 1938, detailing a Martian invasion of New Jersey
5. Earth-based telescopic observations of Mars revealed that its axial tilt and length of day are very similar to those on Earth, and that seasonal variations occur in surface markings and the polar ice caps
6. Many space probes, primarily from the United States, have provided detailed knowledge of Mars

a. *Mariner* probes in the early 1970s revealed that the surface is barren and cratered
   b. The greatest detail came from the U.S. *Viking* missions in 1976, which included orbiters and landers conducting observations and experiments on the surface for several years

B. **The atmosphere of Mars**
   1. Mars has a very thin atmosphere (the pressure at the surface is only 0.006 times the pressure at sea level on Earth)
   2. The Martian atmosphere was once much thicker than it is today, which would have made the surface warmer as a result of the greenhouse effect
   3. Composed chiefly of carbon dioxide (95%), the Martian atmosphere has some nitrogen and argon as well as trace amounts of other gases
   4. Water vapor is minimal, but some water ice is present in the Martian polar caps, which are composed chiefly of frozen CO_2 (dry ice)
   5. Mars has complex and variable winds much of the time, but they have relatively little impact because of the thinness of the atmosphere
   6. Mars has seasons because of the tilt of the planet's axis (just under 24°); the seasonal variations are modified by the relatively noncircular shape of the planet's orbit about the Sun
      a. Mars is farthest from the Sun when it is summer in the northern hemisphere and winter in the southern hemisphere
      b. Mars is closest to the Sun when it is winter in the northern hemisphere and summer in the southern hemisphere
      c. Therefore, the seasonal variations are relatively mild in the northern hemisphere but extreme in the southern hemisphere
      d. Every southern summer, when the southern hemisphere is subjected to intense solar heating, global dust storms are triggered; they sometimes shroud the entire planet with fine dust
      e. The global dust storms alter the dust covering on the surface, and are responsible for the annual surface color variations that were once attributed to changing foliation on seasonal plants

C. **The surface of Mars**
   1. Most of the northern hemisphere of Mars is covered by plains, while a large portion of the southern hemisphere consists of an enormous plateau called Tharsis that covers about 25% of the entire planet
   2. The surface is barren and rocky, with many impact craters
   3. There is a huge valley called Valles Marineris (named after the *Mariner 9* spacecraft, which obtained the first photos revealing the canyon's existence)
      a. This valley is five to 10 times larger than the Grand Canyon in every dimension; it is some 4,500 km long, up to 700 km wide, and 7 km deep
      b. Valles Marineris appears to have formed as a result of crustal cracking when the Tharsis plateau was uplifted
   4. Several large volcanoes also are seen on Mars
      a. The largest, called Olympus Mons, is 27 km tall (over three times the height of Mt. Everest) and has a base 700 km in diameter

        b. These are shield volcanoes similar to those that form the Hawaiian Islands; the reason they are so much bigger on Mars is that there is no continental drift to carry the mountains away from the subsurface volcanic vents that formed them
    5. The evidence for tectonic activity on Mars is minimal
        a. The Tharsis plateau uplift may have been caused in an early period (about 4 billion years ago) when Mars was geologically active
        b. Today there is no widespread system of crustal plates and no evidence of crustal motion similar to continental drift
        c. Scientists dispute how recent volcanic activity may have ceased on Mars; some believe there may have been eruptions within the past billion years; others suggest all the activity stopped more than 3 billion years ago
        d. No magnetic field has been detected, suggesting that there are no extensive molten regions in the planet's interior
    6. Close-up examination of the Martian surface was accomplished by the two *Viking* missions
        a. Photographs of the surface rocks and dust showed that they are quite reddish in color, due to high concentrations of iron oxide; this explains the overall color of the planet
        b. The *Viking* landers were equipped with scoops and test chambers with which the Martian soil could be analyzed directly

D. Water on Mars
    1. Evidence suggests that Mars once had ample supplies of water
        a. Water should have been present from early times; its origins were in the venting of gases from the interior of the planet and from accretion of icy bodies from space (cometary nuclei)
        b. Ancient, dry river channels are seen today on Mars, particularly along the edges of the southern highland regions; flood plains also are evident
        c. Some valleys, including portions of Valles Marineris, show signs of ancient beaches where there once was standing water in the form of large lakes
    2. If there had been liquid water on Mars, the atmosphere must have been much thicker; the current atmosphere on Mars is too thin prevent liquid water from evaporating quickly
    3. Some or all of the water present today may be in the form of subsurface ice
        a. This ice is called *permafrost*, and similar subsurface deposits are found in arctic regions on Earth
        b. There are regions on Mars where it appears that subsurface ice has suddenly melted, causing the surface to collapse; the resulting jumbled areas are called *chaotic terrain*

E. The search for life on Mars
    1. Scientists observing from Earth could not establish evidence of Martian life; they explained the dark surface markings and the seasonal variations in other ways
    2. The *Viking* missions conducted searches for microscopic life-forms in Martian soil samples and found no organic molecules in the soil; although no evidence for life was found, the possibility that life once existed on Mars was not ruled out
    3. The fact that Mars once had liquid water and a denser, warmer atmosphere suggests that primitive life-forms might once have existed there; thus, future missions may be designed to look for fossil microorganisms

F. The Martian moons
1. Phobos and Deimos are tiny and orbit very close to Mars
2. In close-up photos, they are irregular in shape and marked with craters
3. The two tiny moons most likely were captured gravitationally by Mars from the nearby asteroid belt

Study Activities

1. Make scale drawings of the four terrestrial planets, discussing their similarities and contrasts in terms of such general properties as mass, radius, and average density.
2. Summarize the similarities and differences between the Moon and Mercury.
3. Explain how the Sun's tidal force has caused Mercury to become locked into 3:2 spin-orbit coupling.
4. Explain why it is so hot on Venus.
5. Summarize the evidence for the past (possibly present) existence of volcanic activity on Venus.
6. Explain how the surface conditions on Venus and Earth came to be so different despite the general similarity between the two planets.
7. Compare and contrast the atmospheres of Venus, Earth, and Mars; point out which features are similar and which are different.
8. Summarize the evidence that water once existed on Mars.

9

The Giant Planets and Pluto

Objectives

After studying this chapter, the reader should be able to:
- Describe the general properties of each of the outer five planets.
- Describe and explain the wind patterns in the atmospheres of Jupiter, Saturn, Uranus, and Neptune.
- Summarize the composition of the atmospheres of the four giant planets, and point out their differences and similarities.
- Describe the interior structure of the giant planets, and explain the internal similarities and differences among these planets.
- Discuss the four Galilean satellites of Jupiter, explaining their differing levels of geological activity.
- Describe the ring system of Saturn, explaining how the complex structure of the rings is maintained.
- Discuss and compare the "twin planets," Uranus and Neptune.
- Summarize the unusual nature of Pluto and its satellite.

I. Jupiter

A. General information
1. Jupiter is the fifth planet from the Sun, orbiting at an average distance of about 5.2 times the distance between the Sun and Earth
2. Jupiter is the largest and most massive of all the planets, having a diameter nearly 11 times that of Earth and a mass more than 300 times greater
3. Jupiter's density is far lower than that of the terrestrial planets
   a. The planet's composition is dominated by the lightweight elements, especially hydrogen and helium
   b. Jupiter (and the other giant planets) retained these lightweight elements because of the planet's distance from the center of the solar system; the elements condensed and became trapped
4. Jupiter rotates so rapidly that the planet itself is somewhat flattened, having a 7% greater diameter through its equator than through its poles
5. The rapid rotation is the result of Jupiter's formation (see Chapter 6, Solar System Formation and Planetary Science); as Jupiter was growing by accreting gas from its surroundings, conservation of angular momentum forced its spin rate to increase as it gained matter and contracted
6. From Earth, historical telescopic observations revealed many details about Jupiter
   a. Galileo discovered its four large satellites in 1609

b. The banded structure of the atmosphere, consisting of dark belts and bright zones, is readily seen using Earth-based telescopes
c. The Great Red Spot, a reddish-brown oval structure in Jupiter's southern hemisphere, also is easily seen

7. Scientists have discovered that Jupiter is a powerful emitter of radio waves
    a. Some of the radio emission comes from charged particle belts confined by a strong magnetic field
    b. Lightning discharges in the upper atmosphere of the planet also create radio emissions

8. Jupiter has been visited by several space probes, all from the United States
    a. The *Pioneer 10* and *Pioneer 11* probes flew past Jupiter, taking crude photos and making measurements of local conditions
    b. The *Voyager 1* and *Voyager 2* spacecraft obtained many high-quality color images of Jupiter and its moons
    c. The *Galileo* probe is now on its way toward Jupiter, due to arrive in late 1995, when it will enter Jupiter's orbit and remain there for months

B. The atmosphere and interior of Jupiter

1. The internal structure of Jupiter is determined from a combination of external observation and theoretical analysis
    a. The interior is entirely fluid, except for a small, rocky core of heavy elements that have sunk to the center through differentiation
    b. The outermost layer is gaseous, dominated by hydrogen, helium, and hydrogen compounds (such as ammonia [NH_3] and methane [CH_4])
    c. Below this layer is a deep zone where hydrogen exists in a liquid state
    d. Below the liquid hydrogen layer is a region extending to the core where hydrogen is in a liquid-metallic state, consisting of a semirigid crystalline structure that exists only under conditions of extremely high pressure and temperature
    e. Jupiter emits more energy (in the form of infrared radiation) than it receives from the Sun; the excess radiation is thought to arise from heat trapped inside the planet since its formation some 4.5 billion years ago

2. A combination of convection (resulting from heating from below) and the rapid rotation of the planet creates the complex circulation in the upper atmosphere
    a. Solar heating, an additional source of heat, also helps to drive the convection
    b. The resulting rotary winds are stretched into planet-girdling belts and zones; the belts are dark bands where gas is sinking, while the zones are light bands where the gas is rising

3. The Great Red Spot is a long-lasting storm system
    a. This is the largest of many rotary storm systems on Jupiter, and it has been present for at least 300 years
    b. Theoretical studies show that the Great Red Spot could have formed from turbulent eddies in the normal atmospheric flow of the upper clouds

C. Particle belts and the magnetic field

1. The *Pioneer* and *Voyager* spacecraft made direct measurements of the charged-particle belts that had been discovered earlier by radio observations from Earth
2. The rapid rotation of Jupiter (and its magnetic field) has flattened the particle belts into a thin sheet or disk in the equatorial plane of the planet

Jupiter

3. Detailed measurements of the structure of the magnetic field show that it arises from the deeper layers of Jupiter, where fluid circulation creates the necessary electrical currents

D. The satellites of Jupiter
  1. Jupiter has 16 known moons; four of them are large (collectively known as the Galilean satellites), and the rest are quite small
      a. Some of the small moons were not discovered until the *Voyager* encounters in the late 1970s
      b. The four Galilean satellites, from outermost to innermost, are Callisto, Ganymede, Europa, and Io; some of the small moons are closer to Jupiter than Io, while others orbit beyond Callisto
  2. Callisto has an ancient surface, dominated by impact craters
      a. The surface is made largely of ice, which is darkened because of long exposure to ultraviolet radiation from the Sun
      b. The impact craters are white because shattered ice is reflective
  3. Ganymede is the largest satellite in the solar system
      a. The surface shows many impact craters, so it is quite old
      b. Grooved regions indicate some past tectonic activity
  4. Europa has a cracked surface, lending it the appearance of a crystal ball
      a. The few impact craters on Europa suggest that the surface is young; that is, it has been renewed or modified relatively recently
      b. The dark linear features appear to be ice-filled cracks in the crust
  5. Io is geologically active
      a. No impact craters appear on Io, indicating that the surface of this satellite is very young
      b. The brilliant yellow and orange colors are typical of recent, sulfurous volcanic deposits
      c. The *Voyager* probes discovered actively erupting volcanoes at many locations on Io
      d. The volcanic activity, combined with the measured density and temperature of Io, indicate that its interior is almost completely molten
      e. Gas ejected by the eruptions becomes ionized and then is forced by Jupiter's magnetic field to travel around the planet
          (1) This ring of ionized gas following Io's orbit is called the *Io torus*
          (2) The torus wobbles up and down because of the small misalignment (11°) between the equatorial plane of Jupiter and its magnetic equator
  6. Tidal forces explain the varying degrees of geological activity among the Galilean satellites
      a. Jupiter's tidal force has locked each satellite into synchronous rotation, with one side always facing Jupiter
      b. The tidal force is strongest for the innermost satellites; thus, Io and Europa are subjected to the greatest force
      c. Adding to the stresses caused by Jupiter's tidal force, Io and Europa regularly exert forces on each other
          (1) The orbital period of Europa is exactly twice that of Io, so the two are aligned regularly on the same side of Jupiter
          (2) A simple ratio of orbital periods, such as the one between Io and Europa, accompanied by frequent alignments is called **orbital resonance**

(3) The frequent alignments of Io and Europa cause enhanced tidal stresses on both, helping to explain Io's volcanic activity and the milder tectonic activity on Europa
   d. Thus, the Galilean satellites experience varying degrees of internal stress caused by external tidal forces, with the greatest stress being exerted on Io and the least on Callisto; this is why Io is so active geologically while Callisto is inert and the other two are somewhat active geologically
   e. The original composition of each of the Galilean moons was approximately half rock and half ice, but the ice has melted and evaporated (this phenomenon depended on the extent of tidal heating); consequently, Io has the least ice and the highest density today, while Callisto has the most ice and the lowest density
   f. The interior of Europa may be warm enough for liquid water to exist underneath the crust; for this reason, some scientists speculate that life could have begun there
7. *Voyager* observations showed that Jupiter has a single, thin, dim ring

II. Saturn

A. General information
1. Saturn is the sixth planet from the Sun, orbiting at an average distance of about 9.5 times the distance between the Sun and Earth
2. Although Saturn is not as large as Jupiter, it has a diameter some 9 times larger than that of Earth and a mass that is 95 times greater
3. The density of Saturn is lower than that of Jupiter; in fact, it is lower than the density of water
4. Saturn's most distinguishing feature is its beautiful system of rings
   a. Galileo was unable to determine the nature of the rings, although his telescope revealed that Saturn has an elongated appearance
   b. In the mid 1600s, Dutch astronomer Christiaan Huygens deduced the nature of the elongated shape of Saturn, realizing that it was due to an equatorial ring

B. Atmosphere and interior
1. Atmospheric circulation, driven by convection and the rapid rotation of the planet, has created a banded appearance similar to that of Jupiter
   a. The belts and zones of Saturn are much less distinct than those of Jupiter because the atmosphere above the clouds is thicker on Saturn
   b. The wind speeds on Saturn are much higher than on Jupiter
   c. The circulation pattern differs between the northern and southern hemispheres on Saturn because, unlike Jupiter, Saturn has a substantial axial tilt (almost 27°) and therefore has seasonal variations
2. The atmosphere of Saturn has numerous spots, similar in appearance to the many small spots on Jupiter
   a. Saturn displays no major features as large or long-lasting as the Great Red Spot of Jupiter
   b. In 1991, Saturn exhibited a "Great White Spot," an equatorial storm system that lasted for several months (not hundreds of years)

3. The composition of Saturn's atmosphere (like that of Jupiter) is dominated by helium and hydrogen compounds
   a. One contrast with Jupiter is that much of the ammonia (NH_3) has precipitated out of Saturn's atmosphere in the form of ammonia snow because of Saturn's colder temperature
   b. The colder temperatures also have allowed the formation of more complex molecular species (such as ethane [C_2H_6]) in Saturn's atmosphere
4. The interior structure of Saturn, like Jupiter, is thought to consist of a liquid hydrogen zone beneath the clouds, a liquid metallic hydrogen zone below that, and a small, solid core containing heavy elements
5. Saturn also emits excess radiation, but in this case the source of heat may be released gravitational potential energy resulting from differentiation that is still occurring (as helium atoms slowly sink toward the center of the planet)
6. The structure of Saturn's magnetic field suggests that it originates at a deeper level in the planet's interior compared to Jupiter's magnetic field

C. The moons of Saturn
1. Saturn has 18 verified moons, with other small ones suspected
   a. The moons fall into three categories: the giant Titan, the seven intermediate-sized moons, and the remaining small moons
   b. *Voyager* discovered most of the small moons; Titan and the intermediate-sized moons were seen from Earth much earlier
2. Titan is unique among satellites in the entire solar system because it has a thick atmosphere (the only other moon that has an atmosphere is Triton of Neptune, which has a thin, ephemeral atmosphere)
   a. Titan's atmosphere is denser and thicker than that of Earth; Titan's atmospheric pressure at the surface is 50% greater than Earth's atmospheric pressure at sea level
   b. The atmosphere of Titan is composed chiefly of nitrogen, like that of Earth, but little or no oxygen is present
   c. At Titan's surface, methane (CH_4) and ethane (C_2H_6) may exist in liquid form; thus, Titan may have liquid lakes or oceans
3. The seven intermediate-sized moons consist primarily of ice
   a. The surfaces of most intermediate-sized moons are heavily cratered, indicating that they have been geologically dormant for a long time
   b. In one or two cases, such as Enceladus and Rhea, there is evidence of surface renewal as water vapor escaped from the interior and froze; as a result, Enceladus has a shiny surface
   c. The satellite Iapetus has a very dark region on one side, which may be caused by the accretion of a sooty substance of unknown origin
   d. Another intermediate-sized moon, Hyperion, has a flattened shape, in contrast with the spherical shapes of all the other intermediate-sized satellites
4. Many of the very small moons orbit within the ring system of Saturn, while others are associated with intermediate-sized moons
   a. Gravitational effects of the tiny "ring moons" help to create some of the intricate structure of the rings
   b. A few of the small moons share orbits with larger ones, at positions 60° ahead or behind the larger one; these are called *co-orbital satellites*

(1) The positions where the co-orbital satellites are found are called *Trojan points;* they are unusually stable as a result of the combined gravitational forces of Saturn and the large satellites
(2) Tethys has two co-orbital satellites, and Dione has one

D. The rings
1. Even from Earth, observations show that the rings are made of small particles and have a complex structure
   a. The rings appear semitransparent
   b. Spectroscopic measurements employing the Doppler effect show that the particles follow individual orbits around Saturn at speeds dictated by Newton's laws of motion
   c. At least three broad rings can be distinguished from Earth in addition to one prominent dark gap
2. Astronomers can deduce the sizes of the rings' ice particles from their light-scattering properties; ring particles range in size from millimeters to tens of meters
3. The rings exist primarily because tidal forces prevent the particles from coalescing to form large moons
   a. Inside a certain distance from Saturn, known as the **Roche limit,** the tidal force acting to pull a satellite apart is greater than the gravitational force acting to hold it together
   b. Nongravitational forces can hold a small body together despite the tidal forces; thus, some of the small moons of Saturn that are satellites of larger moons exist inside the Roche limit
4. The gravitational effects of Saturn's satellites contribute to the structure of the rings
   a. Gaps, such as the prominent Cassini division, occur where ring particles would be in orbital resonance with a satellite
      (1) For example, particles at the position of the Cassini division would have exactly one-half the orbital period of the intermediate-sized moon Mimas, and the repeated alignments with Mimas over time would have disturbed any particles at that position to move to other orbits
      (2) Other gaps are found at positions where particles would have orbital periods equal to one-third, one-quarter, and smaller simple fractions of the period of Mimas
   b. A pair of small moons in nearby orbits can trap ring particles between them, helping to maintain a thin ring; such moons are called *shepherd satellites*
   c. Oscillatory motions triggered by the larger moons have established **spiral density waves** in some portions of the ring system
      (1) A spiral density wave is a spiral-shaped standing wave consisting of alternating dense and rarefied regions
      (2) In regions of the ring system where spiral density waves dominate, the rings consist of closely spaced spirals, similar to the grooves on a record
5. The magnetic field also contributes to the rings' structure
   a. Tiny particles in the rings can develop sufficient electrical charge by accumulating free electrons so that magnetic forces, rather than gravity, dominate their motion

  b. Electrical forces may account for the *spokes*, which are dark streaks directed outward from Saturn; these persist for relatively long periods of time
  c. These dark bands are thought to be created by very small particles that are electrically charged and suspended above and below the plane of the rings by electrical forces
6. Current theories suggest that the rings are not stable but instead may alter in shape and structure over time
  a. The asymmetrical shapes of some of the rings, which are unstable gravitationally, are evidence that the rings must change with time
  b. Rings may be replenished from time to time by collisions that destroy satellites
   (1) Such collisions with incoming meteoroids are especially likely close to giant planets whose gravitational influence attracts meteoroids; this phenomenon is called *gravitational focusing*
   (2) Further evidence for occasional destruction of satellites by collisions is found in the ring and moon systems of Uranus and Neptune; both systems exhibit features that are unstable over long periods of time

III. Uranus and Neptune

A. General information
1. Neither Uranus nor Neptune is visible from Earth to the unaided eye; thus, both planets were unknown to ancient astronomers
  a. Uranus had been charted as a star previously, but was not recognized as a planet until English astronomer William Herschel discovered it in 1781
  b. Scientists discovered Neptune following the prediction of its existence (by John C. Adams and Urbain Leverrier) based on its gravitational effect on the orbit of Uranus; it was seen in 1840 by German astronomer Johann Galle
2. Uranus orbits the Sun at an average distance of 19.2 AU; Neptune's average distance from the Sun is 30.1 AU
3. The two are very similar in overall properties, each having a diameter about four times Earth's and a mass between 14 and 17 times greater than Earth's
4. Both planets exhibit a blue color as a result of the absorption and scattering of red light by methane gas in the atmosphere
5. The axis of Uranus is tipped 98°, so that its north pole points 8° below the plane of its orbit
  a. This causes unusual seasons, as one hemisphere and then the other is pointed almost directly toward the Sun
  b. The strange tilt is thought to be the result of a collision with a large planetesimal early in the history of Uranus

B. Atmospheres and interiors of Uranus and Neptune
1. The composition of both atmospheres is dominated by helium and hydrogen compounds, like Jupiter and Saturn
2. Both atmospheres have belts and zones similar to those of Jupiter and Saturn, although those on Neptune are more pronounced than on Uranus
  a. The "normal" circulation of the atmosphere of Uranus was a bit of a surprise because scientists expected that heating of the north polar region by direct

sunlight would alter the flow pattern; apparently the equatorial region is still warmer than the pole, but the reason for this is unknown
- b. The circulation pattern is less dynamic on Uranus than on Neptune, presumably due to the relatively small temperature contrast between the polar region and the equator on Uranus
- c. Neptune, unlike the nearly featureless Uranus, has some spots (including a Great Dark Spot), and there are a few white wispy clouds of crystallized methane

3. Uranus and Neptune have generally similar interiors, with some differences
   - a. Both are thought to have rocky cores and liquid hydrogen zones, perhaps intermixed with some rock, but probably do not have sufficient internal pressure to support liquid metallic hydrogen zones
   - b. Uranus has almost no excess heat radiation, while Neptune has a high quantity of excess internal heat
   - c. Both planets have magnetic fields and radiation belts; both magnetic fields are out of alignment (by about 60° in each case) with the planetary rotation axis

C. Moons and rings

1. Uranus has 15 known moons, 10 of which were unknown before the *Voyager 2* encounter
   - a. Five are moderately large, while the remaining 10 are small moons, orbiting within the ring system of Uranus
   - b. The five larger moons show varying degrees of geological activity; Miranda, the innermost, has a bizarre surface that may be the result of fragmentation and subsequent re-formation of the entire body
2. Uranus has 9 distinct, thin, dim rings
   - a. The rings were detected from Earth in 1977 as they *occulted* (shut off from view) a star beyond the rings, so that the star appeared to dim and brighten alternately as the rings passed in front of it
   - b. The inner portion of the ring system consists of a broad, dark band of dust particles; calculations show that this band will not be long-lasting, indicating that it must be replenished somehow
   - c. Other rings have unstable structures, further indicating that the ring system must change with time
   - d. It has been suggested that collisions occasionally destroy satellites, creating new ring particles (this may explain the strange geology of Miranda)
3. Neptune has 8 known moons, while additional small ones are suspected
   - a. Only two, Triton and Nereid, were known before the *Voyager 2* encounter
   - b. Triton orbits Neptune in the backward, or retrograde direction, relative to most other motions in the solar system
   - c. Triton has a thin atmosphere and exhibits evidence of active venting of gases in the form of ice geysers, which were seen in *Voyager* images
   - d. Nereid orbits Neptune in the normal direction, but has a strangely elongated orbit with a long orbital period (about 1 year)
4. Neptune's rings are not uniform and apparently transient
   - a. Before the *Voyager* encounter, it was thought that the rings did not completely encircle the planet but instead consisted of arcs and fragments of rings

b. The *Voyager* images showed that the rings are complete but lumpy, so that they appear brighter in some places than others
c. The unusual, asymmetrical shape of the rings of Neptune contributes to the recent suggestion that ring systems in general are dynamic, changing entities

IV. Pluto

A. General information
1. Pluto was discovered in 1930 by American astronomer Clyde Tombaugh
   a. The discovery was the result of an intensive search stimulated by the erroneous belief that the motions of Uranus and Neptune were being perturbed by a ninth planet
   b. The discovery was made by careful comparisons of photographs of the same portion of the sky taken at different times, so that the motion of a planetary body would be seen in contrast with the fixed stars
2. Pluto has a very unusual orbit compared with the rest of the planets; it is tilted by 17° with respect to the ecliptic plane, and it is so elongated that Pluto's distance from the Sun varies between 29 and 49 AU
3. Pluto's unusual orbit causes it to fall within Neptune's orbit for 20 years of its 248.5-year orbit around the Sun; this most recently began in 1979 and will end in 1999
4. Pluto is very small and low in mass, having a diameter of less than 20% of Earth's and a mass of only 0.00002% of Earth's
5. In 1977, Pluto was found to have a satellite (named Charon), which is very large relative to Pluto itself
   a. The orbital period of Charon is equal to the rotational period of both Pluto and Charon
   b. Thus, this double planet is in double-synchronous rotation; tidal forces have slowed the rotations of both bodies so that each keeps the same face toward the other at all times

B. Charon's role in understanding Pluto
1. Charon's orbital period and orbital separation from Pluto allowed Pluto's mass to be deduced using Kepler's third law of motion
2. The tilt of Charon's orbital plane, which coincides with Pluto's equatorial plane, allowed the tilt of Pluto's axis to be measured (see *Orientation of Pluto's Spin Axis and Charon's Orbit,* page 82)
   a. Pluto's north pole points 118° away from vertical, or 28° below its orbital plane
   b. Thus, Pluto technically rotates in the retrograde direction, as do Venus and Uranus
3. In the late 1980s, the orbital plane of Charon was aligned in the direction from Pluto to the inner solar system, so that Charon and Pluto repeatedly eclipsed each other as they orbited
   a. The duration of the eclipses allowed the diameters of Pluto and Charon to be measured accurately

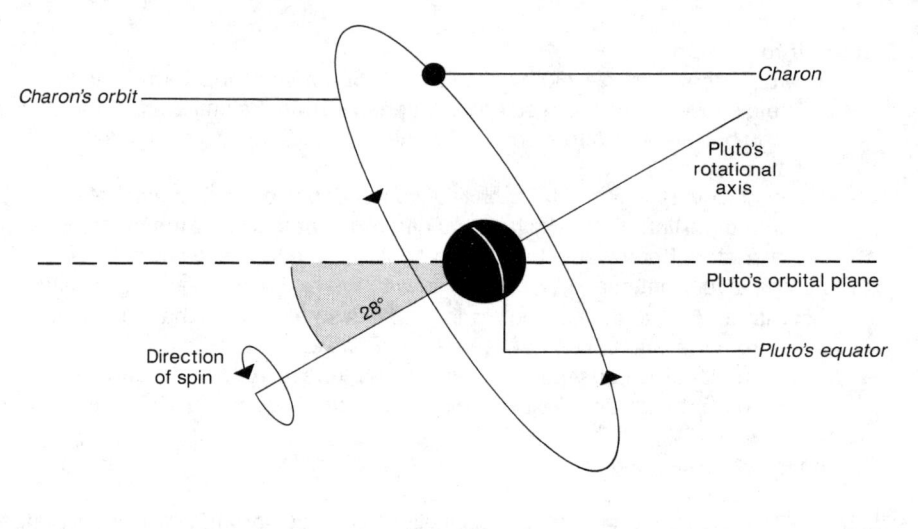

Orientation of Pluto's Spin Axis and Charon's Orbit

Pluto's rotation axis is tipped over so far that it points below the plane of the planet's orbit around the Sun. Relative to the ecliptic, Pluto itself is tipped by 17°, an unusually large angle. Astronomers therefore identify the spin of the planet as retrograde (that is, backward). Orbiting Pluto, the moon Charon is in the ninth planet's equatorial plane. During the late 1980s, astronomers observed that the plane of Charon's orbit was aligned so that Pluto and Charon alternately passed in front of each other with respect to Earth.

    b. Once the diameters and masses were known, scientists could calculate the densities of both bodies, leading to information about their internal compositions (about half rock and half ice)

    c. Scientists derived information about Pluto's thin atmosphere by observing Charon as it passed behind Pluto and was gradually obscured by the ninth planet's atmosphere

C. Atmosphere and interior of Pluto
    1. Spectroscopic studies show that Pluto's atmosphere primarily is composed of methane, with another gas (probably nitrogen) mixed in
    2. Pluto's surface conditions are nearly cold enough to ensure that all atmospheric gases should precipitate out into solid form
    3. Pluto has a gaseous atmosphere only when the planet is near its closest approach to the Sun (such as during the late 20th century)
    4. Pluto's unusual features suggest that it has a different origin than the rest of the planets do
        a. One earlier hypothesis was that Pluto is an escaped moon of Neptune that underwent a near-collision with Triton and was ejected into its own solar orbit; this idea is ruled out by the fact that Pluto is now known to have a satellite of its own

b. The modern interpretation is that Pluto is a large planetesimal formed in the outer solar system, much in the same manner as the giant moons of Jupiter, Saturn, and Neptune
c. Many more bodies in the outer solar system may exist, similar to the giant satellites and Pluto, but have not been discovered

Study Activities

1. Using information from this chapter as well as Chapter 6, discuss the formation of the giant planets. Explain why the giant planets are so different from the terrestrial planets.
2. Explain how the atmospheric circulation on Jupiter and Saturn is similar to and different from that of Earth.
3. Describe the internal structure of Jupiter and Saturn, showing their similarities and contrasts. Discuss the interiors of Uranus and Neptune, and compare them with those of Jupiter and Saturn.
4. Explain the contrasting properties of the four Galilean satellites of Jupiter.
5. Summarize the similarities and contrasts between Saturn and Jupiter in terms of mass, diameter, density, and composition; then compare Saturn and Jupiter to Uranus and Neptune.
6. Explain how the moons of Saturn influence the complex structure of the planet's ring system.
7. Summarize the arguments suggesting that planetary ring systems are dynamic, changing entities instead of permanent, unchanging structures.
8. Describe the ways in which Pluto is different from both the terrestrial and the giant planets, and comment on what these contrasts may say about the origin of Pluto.

10

Interplanetary Bodies

Objectives

After studying this chapter, the reader should be able to:
- Describe asteroids and their properties.
- Explain the formation of asteroids and the role of Jupiter's gravity.
- Summarize the distribution and orbital motions of comets.
- Describe how a comet changes with repeated passages near the Sun.
- Summarize the origins of long-period and periodic comets.
- Discuss the nature and origin of meteors and meteorites.
- Describe the distribution and properties of interplanetary dust.

I. Asteroids

A. General information
1. *Asteroids*, also called *minor planets*, are small bodies that orbit the Sun and are distinct from planets, satellites, or comets
2. The first asteroid, Ceres, was discovered accidentally by G. Piazzi in 1801
3. Others were soon found, and now thousands are catalogued
4. The total mass of all the asteroids is far less than the mass of even a small planet; the mass is estimated at only 0.04% of Earth's mass
5. Most asteroids orbit between Mars and Jupiter (this commonly is called the asteroid belt); Renaissance astronomers had predicted the existence of a planet in that orbital region based on a mathematical progression called Bode's law that described the distances of the known planets from the Sun
   a. Several asteroids have orbits that cross Earth's orbit; these *Apollo asteroids* have the potential to collide with Earth, causing mass destruction
   b. Small groups of asteroids follow Jupiter's orbit at points 60° ahead of or behind Jupiter; these *Trojan asteroids* are analogous to the co-orbital satellites of Saturn
6. New asteroids are most easily detected in time-exposure photographs because their motion creates trailed images
7. The *Galileo* probe has photographed two asteroids, Gaspra and Ida, from close range

B. Properties of asteroids
1. Several techniques are used to determine the sizes and shapes of asteroids
   a. The brightness of an asteroid provides an estimate of its size, which is related to the amount of surface area available to reflect sunlight

(1) The estimated size of an asteroid depends on what is assumed for its *albedo*, the fraction of incident light that its surface reflects
(2) Most asteroids vary in brightness as they spin; this indicates that they are irregular in shape
b. When an asteroid passes in front of a star, the duration of the eclipse (called a *stellar occultation*) yields information on the size of the asteroid
c. Scientists can beam a radar signal at an asteroid and analyze the reflected beam to determine the asteroid's size and shape
d. Asteroids glow at infrared wavelengths; the total infrared luminosity of an asteroid is a measure of its temperature and surface area
e. The result of all these techniques is that asteroids are found to have diameters ranging from a few km to about 1,000 km; the small ones are usually irregular in shape, while the largest tend to be spherical
2. Infrared measurements determine the composition of an asteroid
a. Solid compounds have distinct spectral bands where they emit or absorb light, much as atoms do
(1) These spectral bands are much broader in wavelength than the spectral lines formed by atoms
(2) Most of the spectral features due to solid material in asteroids lie at infrared wavelengths
b. Asteroids are classified according to their composition
(1) *C-class asteroids*, the most common type, have a high carbon content
(2) Other less common classes include metallic asteroids and various types of silicates
c. The almost purely metallic composition of some asteroids indicates that they originated in larger bodies that have undergone differentiation, thereby forming a nickel-iron core
3. Jupiter's gravity creates distinct gaps in the asteroid belt
a. Where an asteroid would have a period that is a simple fraction of Jupiter's orbital period, frequent alignments between Jupiter and that asteroid tug the asteroid out of that orbit over time
b. This is an example of **orbital resonance,** similar to the mechanism by which Saturn's moons create gaps in the ring system
c. The gaps in the asteroid belt are called **Kirkwood's gaps,** after American astronomer Daniel Kirkwood, who first realized how they are formed

C. Origins of asteroids
1. Tidal forces resulting from Jupiter's gravity are responsible for most asteroids
a. Asteroids may be leftover planetesimals that never coalesced to form a planet
b. Jupiter, by far the most massive planet, exerts tidal forces on the asteroids that keep them moving relative to each other (for information on tidal forces see Chapter 3, The Evolution of Modern Astronomy, and Chapter 9, The Giant Planets and Pluto)
2. The motions of asteroids relative to each other are rapid enough to prevent them from sticking together when they collide
a. Instead, these motions cause violent collisions that fragment many large asteroids, thereby populating interplanetary space with meteoroids
b. Metallic asteroids are thought to be remnants of planetesimals that were originally large enough to have undergone differentiation

II. Comets

A. General information
1. Comets are hazy-looking celestial phenomena usually consisting of a bright head and a faint, wispy tail
2. Comets have historically been associated with calamity, particularly during the Dark Ages in Europe
3. Tycho Brahe was the first to show that comets belong to the realm of space
   a. Previously, comets had been thought of as atmospheric phenomena
   b. Brahe used a comet's lack of parallax as measured from different points on Earth to prove that it was too distant to be part of Earth's atmosphere
4. Edmund Halley, an associate of Newton, was the first to realize that some comets return periodically
5. In modern times, comets are observed with large telescopes and have been visited by space probes, but most are still discovered by amateur astronomers using small telescopes

B. Cometary orbits
1. Most comets follow highly elongated orbits about the Sun
   a. In the absence of outside disturbing forces, the orbits of all comets would be ellipses of high eccentricity
   b. Some comets gain speed through gravitational encounters with planets and assume hyperbolic orbits, meaning that they escape the Sun's gravitational bond after one close passage
   c. Comets only form their glowing, gaseous tails and comas during their visits to the inner solar system, which is a very small fraction of the total orbital period
2. Comets are classified as **periodic comets** or **long-period comets**
   a. Periodic comets return to the inner solar system repeatedly, with periods ranging from a few months to several decades (as in the case of Comet Halley, which has a 76-year period) and centuries
   b. Some long-period comets do not return at all after their initial passage because they attain escape speed, while others have periods so long (millions of years) that for all practical purposes they are only seen once
   c. The orbits of periodic comets tend to lie in or near the plane of the planetary orbits, while the orbits of long-period comets are randomly oriented to the plane of the solar system
3. The studies of Dutch astronomer Jan Oort suggested that long-period comets originate in a huge spherical cloud of icy bodies centered on the Sun and have a radius of about 100,000 AU
   a. This reservoir of cometary bodies is known as the **Oort cloud**
   b. Astronomers observe new comets when they are disturbed by a gravitational force or a collision with another body, and they fall in toward the center of the solar system
   c. Because these bodies fall from such a great distance, they follow highly elongated (eccentric) orbits with very long periods
4. A recent theory suggests a different origin for the periodic comets
   a. Oort's theory does not explain why periodic comets tend to have orbits lying in the plane of the solar system, if the Oort cloud is spherical

b. The modern view holds that a second reservoir of periodic comets, called the ***Kuiper belt,*** is a disk lying some 30 to 50 AU from the Sun
   (1) Bodies in the Kuiper belt may be detectable from Earth, unlike bodies in the more distant Oort cloud
   (2) Two recently detected nonplanetary objects beyond the orbit of Uranus may be members of the Kuiper belt

C. Properties of comets
1. A comet consists of a small, solid *nucleus* and, when it is near the Sun, a gaseous portion consisting of a head and a tail
2. The nucleus is a mixture of ice and rocky material
   a. The ice evaporates when heat from the Sun intensifies during the comet's passage through the inner solar system; this forms the head and the tail
   b. The rocky material forms a crust surrounding the nucleus
3. The head of a comet consists of a *coma* and a *halo*
   a. The coma is a roughly spherical volume of glowing gas that stays with the nucleus as it moves in its orbit
      (1) The coma glows by a mechanism called *fluorescence,* in which atoms or molecules gain energy by absorbing ultraviolet photons from the Sun and then lose energy by emitting visible-wavelength photons
      (2) The gas in the coma is mostly water vapor when first released from the nucleus, but quickly breaks apart into atoms and ions
   b. The halo is an extended volume of gas, mostly in the form of hydrogen atoms, and is a fraction of an AU (that is, tens of millions of km) in diameter; the hydrogen is released when ice, the primary constituent of the nucleus, is broken down
      (1) The halo glows in the ultraviolet spectrum, where hydrogen atoms have strong spectral lines
      (2) The gas is primarily composed of hydrogen (along with some oxygen) atoms because the principal gas released from the nucleus is water vapor, which is composed largely of hydrogen (and oxygen)
4. A comet may have two tails
   a. The *ion tail* consists of ionized gas from the nucleus
      (1) The ion tail points nearly straight away from the Sun because of the flow of charged particles from the Sun called the **solar wind**
      (2) It may actually precede the comet as it moves away from the Sun
   b. The *dust tail* consists of tiny solid particles released from the nucleus
      (1) The dust tail tends to trail the comet but also is directed away from the Sun by **radiation pressure** (the impact of photons of light on the dust particles)
      (2) The net result is a smooth, curving tail, typically quite distinct from the ion tail

D. The physical evolution of a comet
1. Comets are among the most primitive bodies in the solar system because they are thought to be leftover planetesimals that have never been subjected to significant heating
   a. Oort cloud comets were probably thrown out to that great distance (around 100,000 AU) when the young Sun was ejecting matter in an unstable early stage

b. Gravitational encounters with giant planets may have propelled Kuiper belt comets to their distances early in the history of the solar system
2. Comets lose some of their material from evaporation of ice each time they pass near the Sun
   a. Scientists estimated that Halley's comet was losing 50 tons of water vapor per second when it was near the Sun during 1985 to 1986
   b. The loss of gas occurs through nozzle-like jets that form in the crust of the nucleus
      (1) Observations of Halley's comet show that the same jets are activated on separate visits to the inner solar system
      (2) The pressure created by the release of the gas from the jets can be sufficient to slightly alter the orbit of a comet
   c. A periodic comet can lose so much icy material from repeated passages near the Sun that it eventually disintegrates
      (1) Several comets have been observed to break apart, being intact on one visit but returning the next time in pieces
      (2) The debris from a fragmented comet is strewn along its orbit, providing a spectacular meteor shower when Earth passes through the trail of debris

III. Meteors and Meteorites

A. General information
1. The term *meteor* refers to the streak of light that is seen when a small solid particle enters Earth's atmosphere from space; the particle itself is called a **meteoroid** and, if any of it survives the descent, it is called a **meteorite**
2. Occasional very bright meteors are called *fireballs* or *bolides*
3. Meteors are studied through the use of wide-angle cameras that survey the sky and record the trails that are made as meteoroids pass through the atmosphere
4. Properties of meteors also are analyzed from studies of meteorites that are retrieved after falling

B. Meteors
1. Meteors are not difficult to see; on virtually any clear night one or more will appear during an evening's watch
2. *Meteor showers* are periods of unusually frequent meteors that always occur on about the same dates each year
   a. During a shower, all the meteors appear to come from the same point in the sky, called the *radiant*
   b. The highest intensity occurs after midnight, when the observer's position has rotated toward the leading side of Earth as it moves in its orbit
   c. Meteor showers tend to be named after the constellation from which they appear to emanate; the *Perseid* and *Orionid* are examples
3. The particles that cause most meteors are small grains of rock that are completely vaporized by the heat of friction in Earth's upper atmosphere
4. Most of the tiny particles that cause meteors are believed to come from comets

C. Meteorites
1. The objects that cause fireballs or bolides sometimes are large enough to reach the ground intact, thus creating meteorites
2. Meteorites are often indistinguishable from ordinary rock, and many go unrecognized
3. Antarctica is an especially good place to find meteorites because they stand out against the snow and ice, and there are few available native rocks to create confusion
4. Meteorites are classified according to composition
   a. The most common are the *stony meteorites*, which comprise about 93% of all meteorites
   b. Most of the remainder are *iron meteorites* (actually a mixture of iron and nickel), which constitute about 6% of all meteorites
   c. The rest are the rare *stony iron meteorites*, in which stone and metal are combined; they are especially difficult to spot among ordinary rocks, so only a fraction of those that fall to Earth are found
5. Most stony meteorites are *chondrites*, meaning they have small spherical inclusions called chondrules
   a. Chondrules form by rapid cooling, indicating that these meteorites probably formed very early in the solar system
   b. Some chondrites are *carbonaceous chondrites*, which appear never to have been heated to any significant degree; they are among the most primitive materials in the solar system
   c. In some carbonaceous chondrites, organic molecules called amino acids have been found, showing that the precursors to life may have reached Earth long ago
6. Astronomers believe meteorites originated as fragments of asteroids
   a. Jupiter's tidal forces may have caused the collisions responsible for asteroid fragments
   b. Scientists have identified "families" of meteorites, based on composition and location of fall; these appear to have originated from the same collision

D. Major collisions
1. Throughout geological time, Earth has been subjected to many collisions by asteroid-sized bodies
   a. Such collisions were more common in the early solar system, when many planetesimals remained from the formation of the planets
   b. Today, such collisions still are possible; over a sufficiently long period of time, they are inevitable
2. Many ancient impact craters have been identified on Earth
   a. Most craters have been erased by geological forces and erosion; the presence of tectonic activity and an atmosphere explains why Earth is not as heavily cratered as the Moon
   b. One of the best examples of an impact crater is the Barringer Crater, near Winslow, Arizona; it is about 50,000 years old
3. Some scientists hypothesize that a major collision about 65 million years ago was responsible for the extinction of dinosaurs
   a. A layer of clay enriched in the element iridium is found throughout the world, separating the geological strata marking the end of the age of dinosaurs; iridium commonly is found in meteorites

b. A huge crater of the right age has recently been discovered on the shore of the Yucatan Peninsula in Central America
   c. Other scientists are skeptical that a collision was the principal cause of dinosaur extinction

E. Interplanetary dust
   1. A medium of fine dust permeates interplanetary space; this dust is concentrated in the plane of the ecliptic
   2. The dust can be observed in several ways
      a. On a dark night, a faint glow resulting from sunlight scattered by the dust can be seen; this is called ***zodiacal light***
      b. A very faint reflection of the Sun, called the ***gegenschein,*** can sometimes be seen exactly opposite the Sun's direction (that is, on the north-south line at midnight)
      c. Infrared observations reveal the presence of dust, which is cool and therefore glows at infrared wavelengths
   3. Interplanetary dust particles can be studied in the laboratory
      a. Particles have been retrieved by high-altitude aircraft and balloon experiments
      b. They also commonly are found in the bottoms of arctic lakes and the oceans, where they accumulate after falling through the atmosphere; they are identified by their characteristic shapes
      c. Laboratory studies show that the particles are complex conglomerates of minerals representing interstellar dust grains that have been merged and processed during the formation of the solar system

Study Activities

1. Summarize the various techniques used by astronomers to study the sizes, shapes, and compositions of asteroids.
2. Explain why asteroids remain in the form of many small bodies instead of a single planet.
3. Summarize the mechanism by which Kirkwood's gaps are formed, and explain how this is similar to the formation of gaps in the rings of Saturn.
4. Describe the orbits of a typical long-period comet and a typical periodic comet.
5. Explain why comets only glow and develop a tail when they are in the inner solar system.
6. Explain why the long-period and periodic comets are thought to originate in different regions.
7. Summarize the chronology of a typical periodic comet, and explain how it is related to the meteor showers that can be observed from Earth.
8. Discuss the origins of meteors and meteorites.
9. Explain why Earth is not as heavily cratered as the Moon.
10. Summarize the ways in which the interplanetary dust is observed.

11

The Sun

Objectives

After studying this chapter, the reader should be able to:
- Describe the general properties of the Sun.
- Explain how the internal structure of the Sun is known.
- Discuss the generation of energy by nuclear reactions inside the Sun.
- Describe the outer layers of the Sun's atmosphere.
- Summarize the role of the Sun's magnetic field in governing the solar activity cycle.

I. Structure of the Sun

A. General information
1. A gaseous sphere, the Sun glows as a result of energy produced in its core by nuclear reactions
2. The Sun produces most forms of energy available on Earth and throughout the solar system
    a. All organic materials grow with the help of sunlight, so the burning of fossil fuels releases energy originally absorbed from the Sun
    b. The Sun's total energy output per second (or its luminosity) is approximately 4×10^{26} watts, the equivalent of 4×10^{24} 100-watt (W) lightbulbs; of this, about 2×10^{17} watts strike Earth
    c. The *solar constant* refers to the energy density (in W/m²) striking Earth (above the atmosphere); its value is 1,360 W/m²
3. Special filters are required to observe the Sun because of its brightness
4. Astronomers have observed spots on the Sun's surface since Galileo's time
5. Observers established the existence of thin, glowing outer layers above the Sun's visible disk during total solar eclipses
    a. The thin, red layer just outside the visible disk is called the **chromosphere**
    b. Extending farther out is a pale, glowing region called the **corona**
6. Modern astronomers learn about the Sun by making observations at many different wavelengths, from the X-ray to the radio portions of the spectrum

B. General properties of the Sun
1. The Sun is an average star, falling near the mean in terms of mass, luminosity, diameter, and other properties
2. The diameter of the Sun is about 7×10^8 m, or roughly 109 times the diameter of Earth

3. The Sun's mass is much greater than that of all the planets added together; it is approximately 2×10^{30} kg, or about 330,000 times the mass of Earth
4. The surface temperature of the Sun is approximately 5,800 K; because the Sun has no solid surface, surface temperature can be defined in numerous ways
    a. **Effective temperature** is the surface temperature required for a sphere the Sun's size to have the same luminosity as the Sun according to the Stefan-Boltzmann law; the effective temperature of the Sun is 5,780 K
    b. Another method is to measure the wavelength of maximum light emission and use Wien's law (see Chapter 4, Light and the Atom); this yields a value of approximately 6,300 K
    c. **Ionization** in the Sun's outer layers also measures temperature; this yields a value near 6,400 K, which varies according to the depth of the solar layer
5. Hydrogen (73% by mass) dominates the Sun's composition, followed by helium (25%), with all other elements together constituting about 2%
    a. This is the same as the initial composition of the solar nebula from which the planets formed
    b. The Sun's composition also represents the overall composition of the galaxy
    c. The composition of the Sun's core is changing gradually because nuclear reactions are converting hydrogen into helium
6. The Sun rotates differentially, meaning that its rotation period at the equator is shorter than at the poles
    a. Only a gaseous or fluid body can rotate differentially
    b. At the equator the rotational period is about 25 days; near the poles it is longer than 28 days
7. The Sun has a strong magnetic field
    a. The field is *bipolar,* meaning that it has the same structure as a bar magnet, with a single north and a single south pole, and magnetic field lines that connect the two
    b. The magnetic field of the Sun is responsible for the structure and motions of the gas in the Sun's outer layers (see below)

C. Internal structure
1. The Sun's internal conditions are derived from observations and theory
2. Temperature, density, and pressure increase sharply toward the Sun's center
    a. The central temperature is about 15 million K
    b. The density at the center is about 160 g/cm^3, or 160 times that of water
    c. The pressure at the center is about 100 billion times Earth's atmospheric pressure at sea level
3. The extremely high temperature in the Sun's interior allows it to remain gaseous despite the fact that the density in the core is far greater than that of lead
4. Additional information about the solar interior comes from analyses of surface oscillations
    a. The surface oscillates because of internal pressure waves that travel throughout the interior
    b. The period of oscillation reveals information about the increase in density from the surface to the core inside the Sun and about rotational speed of the Sun's interior
        (1) A worldwide network of solar telescopes now monitors the Sun continuously as Earth spins

       (2) Observations made from the south pole during the southern summer also have been helpful because the Sun never sets for 6 months at the south pole

D. Energy generation and transport
1. Nuclear fusion reactions generate energy in the Sun's core region
    a. During fusion, nuclei of lightweight elements merge together to form heavier elements
    b. The dominant source of energy in the Sun is a sequence of reactions called the **proton-proton chain** (see *Proton-Proton Chain,* page 94); the net result is that four hydrogen nuclei (protons) merge to form a helium nucleus (two protons and two neutrons)
        (1) The composition of the Sun's core is gradually being converted from mostly hydrogen with some helium to nearly pure helium
        (2) Once all hydrogen is converted to helium after several billion years, other nuclear reactions may take place; thus, we may expect the Sun's core composition to change in the future
    c. In a fusion reaction, the mass of the product is slightly smaller than the sum of the masses of the ingredient particles
        (1) The missing mass is converted into energy according to the formula of relativity: $E = mc^2$
        (2) The mass of one helium nucleus is smaller by 0.007 than the total mass of the four merged hydrogen nuclei; thus, 0.007 of the original mass is converted into energy in the proton-proton chain
    d. Reactions take place only in the solar core because extremely high temperature and pressure are required to make the hydrogen nuclei react
        (1) The nuclei have positive electrical charges and are repelled from each other by electrical forces
        (2) In order to come close enough together to react, the nuclei must collide at very high speeds
2. **Neutrinos,** which are thought to be produced during the proton-proton chain, have not been observed in the expected numbers, thereby creating a current mystery for astronomers
    a. Neutrinos are subatomic particles having no electrical charge and no mass
        (1) The existence of neutrinos was originally suggested from theoretical considerations, but they have been detected experimentally
        (2) Neutrinos do not interact readily with other matter and can escape directly from the Sun's interior into space
    b. Experiments to detect solar neutrinos on Earth have found about one-third of the expected number, indicating that our understanding of neutrinos or the nuclear reactions in the Sun is faulty
    c. New experiments will be completed soon to detect the neutrinos formed in the principal steps of the proton-proton chain; these experiments should provide a more definitive test
3. The conversion of hydrogen to helium in the core will determine the lifetime of the Sun in its present state
    a. This is equal to the total energy available from reactions divided by the rate at which energy is spent; the lifetime (T) equals E/L, where E is the total available energy, and L is the rate of energy loss (which is the luminosity of the Sun)

Proton-Proton Chain

Scientists rely on a short-hand notation to describe nuclei and to express the steps in nuclear reactions. In this notation, a letter or pair of letters represents the chemical element; for example, C stands for carbon, Fe for iron, and so on. Preceding the letter is a superscript to indicate the total number of nucleons (protons and neutrons) in the nucleus. Hence, the common form of carbon, which has 6 protons and 6 neutrons, is ^{12}C. Also preceding the chemical symbol, a subscript indicates the **atomic number**, which is the number of protons. Thus the normal form of carbon, fully described, is shown as $^{12}_{6}$C.

Isotopes are other forms of a given element having the same atomic number but a different number of neutrons. For example, a radioactive form of carbon has 6 protons and 7 neutrons, indicated as $^{13}_{7}$C.

In the proton-proton chain, two hydrogen nuclei (protons, designated as $^{1}_{1}$H) first merge to form a nucleus of an isotope of hydrogen called deuterium, which has a proton and a neutron and is designated as $^{2}_{1}$H. Two subatomic particles, a positron and a neutrino, are then released along with a quantity of energy. The *positron* (e^+) has the mass of an electron and a positive electrical charge, while the *neutrino* (ν) has no mass and no electrical charge. Here is the first step:

$$^{1}_{1}H + ^{1}_{1}H \rightarrow ^{2}_{1}H + e^+ + \nu$$

In the next step, the deuterium nucleus merges with another hydrogen nucleus (proton) to create a form of helium having two protons and one neutron ($^{3}_{2}$He). A gamma-ray photon (γ) is released as well:

$$^{2}_{1}H + ^{1}_{1}H \rightarrow ^{3}_{2}He + \gamma$$

Finally, two $^{3}_{2}$He nuclei merge to form a nucleus of normal helium (2 protons and 2 neutrons, or $^{4}_{2}$He), releasing two hydrogen nuclei. Physicists write the reaction sequence as follows:

$$^{3}_{2}He + ^{3}_{2}He \rightarrow ^{4}_{2}He + ^{1}_{1}H + ^{1}_{1}H$$

In all, 6 protons (hydrogen nuclei) go into the reaction sequence (recall that the first two steps had to happen twice, because two $^{3}_{2}$He nuclei were needed for the final step), and in the end one $^{4}_{2}$He nucleus is formed while two of the protons are returned to their original form. The net result is that 4 hydrogen nuclei merged to form a single $^{4}_{2}$He nucleus, and energy is released.

---

 b. The total energy available is given by $E = mc^2$, where m is the total mass that is converted into energy; for the proton-proton chain, this is 0.007 of the initial hydrogen mass of the Sun's core, which constitutes about 10% of the Sun's total mass, resulting in the value:

  $E = (0.007)(0.1 \text{ solar mass})(2 \times 10^{30} \text{ kg})(3 \times 10^8 \text{ m/sec})^2$

  $= 1.3 \times 10^{44}$ joules

 c. The Sun's luminosity is approximately 4×10^{26} watts (which equals 4×10^{26} joules/sec); thus the lifetime is:

$$T = \frac{(1.3 \times 10^{44} \text{ joules})}{(4 \times 10^{26} \text{ joules/sec})}$$

  $= 3.3 \times 10^{17}$ sec $= 1 \times 10^{10}$ (or 10 billion) years

 d. The Sun's current age is nearly 5 billion years, so it will continue in its present state for another 5 billion years or so

4. Energy transport inside the Sun can occur by convection or radiative transport
 a. Throughout most of the Sun's interior, radiative transport is predominant
  (1) In *radiative transport,* energy is carried by photons of radiation that are emitted initially from the Sun's core

(2) As these photons move outward, they are absorbed and re-emitted (in random directions) repeatedly; it takes several hundred thousand years for a given photon to make its way to the surface
(3) The photons begin as gamma rays at the solar core and are primarily in the form of visible light when they reach the surface
b. Convection is the dominant energy transport mechanism in the Sun's outer layers
(1) **Convection** is the transport of heat energy by moving volumes of hot gas
(2) In the Sun's outer layers, hot gas rises, cools, and returns to the interior, creating vertical circulation; this causes a mottled appearance in the solar surface called *granulation*

II. The Sun's Atmosphere

A. General information
1. The atmosphere refers to the outermost layers of the Sun, measured as far as we can observe (or conversely, the layers from which photons can escape directly into space)
2. The depth to which we can observe the Sun's atmosphere depends on wavelength; images of the Sun made at different wavelengths (such as visible light, infrared radiation, ultraviolet radiation, or X-rays) reveal information about different levels of the solar atmosphere
3. The atmosphere is divided into distinct layers or zones, based on observational distinctions
   a. The **photosphere** is the yellow-white "surface" of the Sun as seen at visible wavelengths
   b. The **chromosphere** is a thin layer just above the photosphere that glows red
      (1) The chromosphere can be seen at visible wavelengths only during a total solar eclipse
      (2) The chromosphere produces several strong emission lines at ultraviolet wavelengths; these can be observed at any time
   c. The corona is an extended region of very thin gas above the chromosphere
      (1) The corona normally is seen only during total solar eclipses
      (2) The corona produces strong emission lines in the ultraviolet spectrum and glows brightly at X-ray wavelengths

B. The photosphere
1. The visible light that escapes the Sun into space arises primarily from the photosphere
2. The density of the photosphere is approximately 1×10^{17} particles/cm^3, or about 1% of the density of Earth's atmosphere at sea level
3. The temperature of the photosphere is near 6,000 K; temperature and density vary according to the depth of the Sun's layers
4. Detailed images of the photosphere reveal a mottled appearance known as granulation
   a. The granules are regions where hot gas is rising from the interior as a result of convection
   b. The detailed structure changes with time as granules appear and disappear

C. The chromosphere
1. Unlike the photosphere, temperature in the chromosphere rises rather than decreases with height; thus, the chromosphere is more highly ionized than the photosphere
   a. Because of its low density and the fact that it is hotter than the underlying photosphere, the chromosphere forms emission lines (see Kirchhoff's rules in Chapter 4, Light and the Atom); one is the red line that characterizes hydrogen
   b. This red line gives the chromosphere its red appearance when viewed during a total solar eclipse
   c. A *spectroheliogram* measures the chromosphere by using a special filter allowing only the light of this emission line to pass through
2. Astronomers believe the elevated temperature in the chromosphere is indirectly caused by the energy of overturning gas in the Sun's outer layers where convection occurs
3. The chromosphere's structure includes a large-scale cellular appearance called *supergranulation* and outward-pointed spikes of glowing gas called *spicules*

D. The corona
1. The temperature continues to rise steeply outward through the corona, reaching a value of between 1 and 2 million K
   a. The cause of this heating may originate in the convective motions in the Sun's outer layers
   b. The corona is so hot that it continuously emits radiation in the X-ray portion of the spectrum (see Wien's law in Chapter 4, Light and the Atom)
   c. The corona also forms emission lines, primarily at ultraviolet and X-ray wavelengths
2. The corona is best observed using an X-ray telescope
   a. X-ray images show large-scale variations in coronal density; low-density regions are called **coronal holes**
   b. The overall structure of the corona changes with time; this is seen when eclipse images are compared

E. Solar wind
1. Early scientific satellites orbiting Earth in the 1960s discovered a stream of charged particles flowing outward from the Sun, now known as the **solar wind**
2. These particles, primarily electrons and protons, stream past Earth at about 300 km/sec
   a. The solar wind distorts Earth's magnetic field, compressing it on the side toward the Sun and extending it in the opposite, or downstream, direction
   b. The flow is known to extend outward through the solar system at least as far as Neptune's orbit
3. The outflow primarily originates in the coronal holes, the regions of low density in the corona, because of the high temperature in the corona, which causes gas particles to escape the Sun's gravitational field
   a. The solar wind is constrained by the Sun's magnetic field to flow outward through the coronal holes
   b. The solar magnetic field rotates with the Sun, causing the solar wind to flow outward in a spiral pattern, much like the pattern of water drops from a rotating lawn sprinkler

III. Solar Magnetic Field and the Activity Cycle

A. General information
1. It has been known for centuries that the Sun has dark spots on its surface, called *sunspots,* which vary in number with time
2. The spots are not actually dark; being cooler than the surrounding photosphere, they are simply less bright and therefore look dark in comparison
    a. Sunspots typically have surface temperatures around 4,000 K, compared to the approximate 6,000 K temperature of their surroundings
    b. Using the Stephan-Boltzmann law, which states that the total energy emitted per second from a thermal emitter is proportional to the fourth power of the surface temperature, we see that a typical sunspot is about one-fifth — $(4,000/6,000)^4$ — as bright as its surroundings

B. Sunspots and the magnetic field
1. Measurements of spectral line analyses reveal that the magnetic fields in sunspots are much higher than in the surrounding photosphere
2. The magnetic field exerts a form of pressure that supports the sunspot gas against the pressure of the hotter photosphere, thereby allowing a sunspot to survive despite its lower temperature
3. Modern models of sunspots view them as locations where magnetic lines of force break through the solar surface
    a. Sunspots typically occur in pairs having opposite polarity, that is, opposite orientations of their north and south magnetic poles
    b. At any given time, all sunspot pairs in the Sun's northern hemisphere have the same orientation of magnetic fields, while those in the southern hemisphere have the opposite orientation
    c. The sunspot pairs may be locations where tubes or loops of magnetic field lines break through the solar surface and then re-enter it; thus, these magnetic structures form loops or arches above the solar surface, with sunspots of opposing polarity as the anchors of the loops or arches
4. Sunspot locations are associated with other forms of solar activity
    a. *Prominences* are large-scale eruptions of gas from the solar surface
        (1) Prominences generally occur at the sites of sunspots
        (2) The erupting gas in a prominence typically follows a looplike or archlike structure that is governed by the solar magnetic field
    b. *Flares* are immensely energetic, explosive outbursts of charged particles
        (1) Flares always arise from sunspot locations
        (2) Flares introduce huge quantities of electrons and protons into the solar wind
        (3) Following a strong solar flare, Earth may be immersed in a strongly-enhanced solar wind about 3 days later (the time required for the wind to flow from the Sun to Earth), thereby disrupting radio communications and creating displays of aurorae
5. Scientists identify these various areas of sunspot activity as *solar active regions*

C. The sunspot cycle
1. Scientists in the 19th century noticed that the number of sunspots varies in a regular cycle, which lasts about 11 years

a. The locations of the spots also vary with this cycle, initially arising well north and south of the solar equator early in the cycle, and then migrating closer to the equator
   b. The number of spots from one 11-year cycle to the next varies, indicating that there may be longer-term effects in addition to the 11-year cycle
   c. A diagram of sunspot patterns over time, covering several successive 11-year cycles, is known as a **butterfly diagram**
 2. In subsequent 11-year cycles, the polarities of the sunspot pairs in each hemisphere of the Sun are reversed; if the northern hemisphere has the eastward spot in each pair with north polarity in one cycle, the eastward spot will have south polarity in the next cycle, and so on
   a. These reversals in sunspot polarity are synchronized with reversals in the Sun's overall magnetic field; the Sun actually flips its magnetic field from pole to pole every 11 years
   b. Thus the 11-year cycle of sunspots is actually a 22-year cycle, because it takes 22 years for the pattern to repeat itself fully
   c. It is not known why the Sun's magnetic field reverses itself periodically, but scientists believe it is related to changes in the circulation of gas in the Sun's interior where the magnetic field arises
 3. Major variations in the solar activity cycle sometimes take place over much longer periods of time
   a. In the late 1600s, the sunspot cycle nearly disappeared, as few or no spots were observed for a prolonged period
   b. This period is known as the *Maunder Minimum,* named after British scientist Maunder, who discovered it through the analysis of historical records (he also invented the butterfly diagram)
   c. During the Maunder Minimum, Earth experienced severe weather and very cool summers; in Europe this was known as the "Little Ice Age"

D. Solar activity cycle and climate on Earth
 1. Scientists have been unable to find a direct link between the sunspot cycle and climate variations on Earth
 2. One complication is that weather patterns on Earth are very complex, making it difficult to recognize cyclical effects
 3. Recent studies in Europe suggest that a relationship between the solar activity cycle and climate exists, but it is not yet well established

---

Study Activities

1. Explain which forms of energy used on Earth originate from the Sun.
2. Explain why nuclear fusion reactions occur only in the central core of the Sun and not throughout the interior.
3. Summarize the contrast between convection and radiative transport of energy. Can you think of everyday examples of each?
4. Calculate what the Sun's hydrogen-burning lifetime would be if its mass were 10 times greater and its luminosity 1,000 times greater than the actual values. What if the Sun's mass were only 1/10 of its actual value, and its luminosity only 1/1,000 of the true value?
5. Summarize the role of the solar magnetic field in forming sunspots.

12

Measuring the Stars

Objectives

After studying this chapter, the reader should be able to:
- Give the reasons and describe the methods for making measurements of stellar positions.
- Summarize the techniques for measuring stellar brightness.
- Use the stellar magnitude system for expressing the brightness of stars and other objects.
- Explain how stellar spectra are recorded, and summarize the stellar spectral classification system.
- Describe the Hertzsprung-Russell diagram and summarize its significance.
- Explain how scientists measure stellar properties, such as mass, luminosity, radius, surface temperature, and composition.

I. Observational Techniques

A. General information
1. Everything we know about stars comes from the analysis of the light we receive from them
2. Most observations can be categorized as one of three general types
   a. **Astrometry** is the measurement of stellar positions
   b. **Photometry** is the measurement of stellar brightness
   c. **Spectroscopy** is the measurement of stellar spectra
3. Scientists commonly require all three types of measurements, as well as information about several different wavelength regions, to analyze a star fully

B. Astrometry
1. Modern astronomers carry out positional measurements for several reasons
   a. Repeated positional measurements can reveal the motions of stars
      (1) The motion of a star (called *proper motion*) in the sky is measured in angular units
      (2) Measurements of stellar motions were instrumental in helping astronomers understand the structure of the Milky Way galaxy (see Chapter 15, The Milky Way) and the properties of star clusters
   b. Positional measurements also establish the distances among stars
2. The most common method of astrometric measurement is to obtain photographic plates and measure star positions relative to each other

a. This requires very high precision and repeatability, both in obtaining the photographs at the telescope and in measuring the positions of star images
   b. The measured relative positions of stars on a plate are converted into absolute positions in the sky by referring to an established set of standard stars whose positions are agreed-upon reference points; thus, reference stars must be included in the photograph
   c. Relative positions of closely spaced stars can be measured using **interferometry** (described in Chapter 4, Light and the Atom)
      (1) In this application, the interference of light rays from adjacent stars is used to determine their angular separation
      (2) Interferometry can be applied only to relatively bright stars
3. The accuracy of typical measurements using these methods is about 0.01 arcsecond
4. New methods of astrometry are being explored
   a. *Hipparcos,* a space-based observatory, takes advantage of the lack of atmospheric blurring; it has achieved an accuracy of about 0.001 arcsecond for thousands of stars
   b. Interferometry instruments in space could measure relative positions to an accuracy of a few millionths of an arcsecond; such instruments are currently under study
5. As Earth orbits the Sun, the position of a nearby star appears to shift slightly; this shifting is called **stellar parallax**
   a. Even the closest stars undergo small parallax motions, too small to have been detected by ancient astronomers without sophisticated instruments
      (1) The closest star, Alpha Centauri, has a parallax motion of only about 1.4 arcsecond during the course of a year
      (2) Stellar parallax was not detected successfully until 1836, when three astronomers achieved it independently
   b. Because the amount of parallax motion a star undergoes depends on how far away it is, stellar distances can be determined by measuring stellar parallaxes
      (1) The *parallax angle* is defined as one-half of the full angular motion that a star undergoes during a year, as Earth orbits the Sun; thus, the parallax angle is the angular displacement of a star as seen from the ends of a baseline 1 AU in length
      (2) By definition, the distance to a star having a parallax angle of 1 arcsecond is one **parsec** (abbreviated pc); 1 pc is equal to 206,265 AU, or 3.26 light-years (see *Stellar Parallax and the Parsec*)
      (3) A star's distance in parsecs is given by the equation $d = 1/p$, where d is the distance and p is the parallax angle; for example, a star having a parallax angle of 0.33 arcsecond has a distance of 1/0.33, or 3 pc
      (4) Because stellar positions generally are not measured to accuracies better than 0.01 arcsecond, stellar distances can be measured only to about 100 pc ($d = 1/0.01$) using this method; this is tiny compared to the size of our galaxy

C. Photometry
1. Originally, stellar brightness was recorded as a means of distinguishing and identifying stars

Stellar Parallax and the Parsec

To illustrate stellar parallax, astronomers can construct a right triangle (a triangle having one 90° angle) with a base of 1 AU, as shown in the diagram below. The parallax angle (p) is the small angle at the peak of the triangle, which is one-half of the full apparent motion the star goes through in a year. If astronomers know two angles of a triangle, they can deduce the ratios of all three sides to each other. If the peak angle is 1 arcsecond (arcsec) in a right triangle, then the ratio of the two legs is 206,265. Thus if the parallax angle of a certain star is 1 arcsecond, then the distance to that star is 206,265 AU. This distance is known as the parsec (pc), a word derived from a contraction of "parallax" and "second." It turns out that 1 pc = 3.26 light-years.

To convert a parallax angle to a distance (d) in parsecs, use the formula d = 1/p. Thus, if a star has a parallax angle of 0.05 arcsec, then its distance is 20 pc (d = 1/0.05).

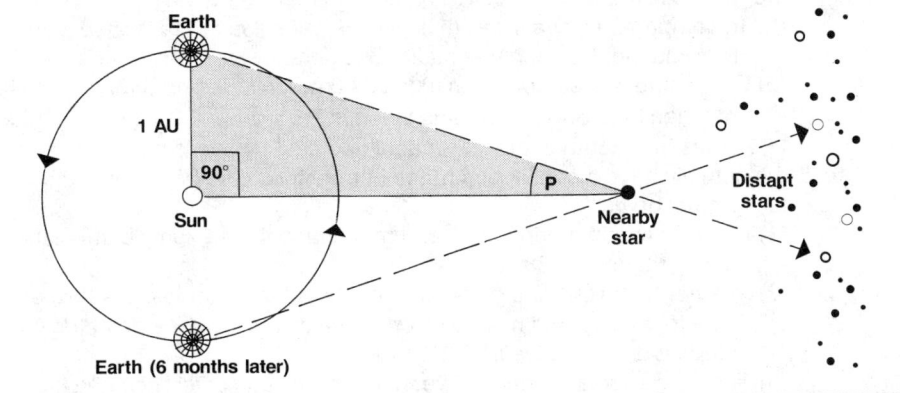

2. Today scientists measure brightness to determine energy output and other physical properties of stars and other objects
3. The most common method of measuring brightness involves the use of a device called a photometer
    a. A photometer usually consists of a vacuum tube, called a photocell, containing a surface that produces an electrical current when light strikes it
    b. The current is proportional to the intensity of the light striking the photocell; thus, measurement of the current is used to measure the intensity of the incident light
    c. Another method is to measure the sizes of star images on photographic plates; the size of an image is related to the intensity of light from the star that formed the image
    d. Astronomers also use electronic detectors that directly record the intensity of incident light
4. Stellar brightness is expressed in terms of the **stellar magnitude system**
    a. The Greek astronomer Hipparchus, in the 2nd century B.C., first used the system of stellar magnitudes
        (1) Hipparchus ranked stars according to their brightness, from first magnitude for the brightest stars to sixth magnitude for the faintest visible to the naked eye
        (2) The system of Hipparchus was imprecise, because it depended on subjective estimates of star brightness
    b. The modern magnitude system is based on precise photometric observations rather than rough estimates

c. The human eye has a logarithmic response to brightness, meaning that star pairs appearing to have similar differences in brightness actually have similar ratios of brightness
   (1) Measurements show that a difference of one magnitude corresponds to a brightness ratio of about 2.5; thus, a first-magnitude star is 2.5 times brighter than a second-magnitude star, a second-magnitude star is about 2.5 times brighter than a third-magnitude star, and so on
   (2) A first-magnitude star is approximately 100 times brighter than a sixth-magnitude star
d. In the modern system, a difference of five magnitudes corresponds to a brightness ratio of 100; thus, the ratio corresponding to a difference of one magnitude is the fifth root of 100, or approximately 2.512
   (1) To compare any two stars, multiply 2.5 times itself (a reasonable approximation) for each magnitude of difference
   (2) For example, a star of magnitude 2 is 3 magnitudes brighter than a star of magnitude 5, which means that it is $2.5 \times 2.5 \times 2.5 = (2.5)^3$ or 16.25 times brighter than the fifth-magnitude star
e. Stars actually have a continuous range of brightness, so fractional magnitudes must be used
   (1) A star between magnitude 3 and magnitude 4, for example, might have a magnitude of 3.68
   (2) In order to measure and compare fractional magnitudes, it is necessary to use logarithms or a scientific calculator (see *Understanding Magnitudes in Mathematical Terms*)
   (3) The precise relationship between the magnitude difference and the brightness ratio is $m_1 - m_2 = 2.5 \log(b_2/b_1)$, where m_1 and m_2 are the magnitudes of two stars and b_1 and b_2 are their brightness (their observed intensities in energy units)
f. Modern measurements have extended the range of stellar magnitudes beyond the original range of magnitude 1 to magnitude 6
   (1) The original first magnitude category actually covers a broad range of brightness; to accommodate this, scientists assign stars magnitudes smaller than 1
   (2) Thus, Sirius, the brightest star in the sky, has magnitude of -1.4
   (3) Telescopes allow stars fainter than magnitude 6 to be observed and measured, so the system extends to magnitudes greater than 6
   (4) The faintest objects observable today have magnitudes of nearly 31; this is 25 magnitudes, or a brightness factor of $(2.5)25 = 8.8 \times 10^9$, fainter than the faintest star visible to the naked eye

D. Spectroscopy
1. The greatest amount of information about a star is derived from analysis of the spectrum of its emitted light
   a. The analysis of ionization and gas motions via the Doppler effect reveals such physical conditions as temperature, density, and turbulence
   b. Measurement of spectral lines helps determine the composition of a star's outer layers
   c. The Doppler effect of a star's spectral lines yields information on the motion of the star along the line of sight

Understanding Magnitudes in Mathematical Terms

Stellar magnitudes are expressed in terms of brightness ratios, and students of astronomy can use "round" numbers to deal with simple multiples of 2.5 (the approximate ratio corresponding to a magnitude difference of 1). However, professional astronomers calculate brightness with the more precise mathematical relationship:

$$m_1 - m_2 = 2.5 \log(b_2/b_1)$$

where m_1 and m_2 are the magnitudes of two stars, and b_1 and b_2 are their brightnesses (expressed in units of energy received on Earth).

The notation "$\log(b_2/b_1)$" means the logarithm of this ratio; a logarithm is the power to which 10 must be raised to give this ratio. For example, if $b_2/b_1 = 10{,}000$, then $\log(b_2/b_1) = 4$, because $10{,}000 = 1 \times 10^4$. Or, if $b_2/b_1 = 0.01$, then $\log(b_2/b_1) = -2$ because $0.01 = 1 \times 10^{-2}$.

Although these are very simple examples, the method works equally well in more complex instances. For example, suppose one star is found to be precisely 142.78 times brighter than another ($b_2/b_1 = 142.78$). Using a scientific calculator (or a logarithm table), $\log(142.78)$ equals 2.155. Substituting for the magnitude difference between the two stars, we find:

$$m_1 - m_2 = 2.5 \log(b_2/b_1) = 2.5(2.155) = 5.39$$

Star 1 is 5.39 magnitudes fainter than star 2 (don't forget that the magnitude scale is backwards: the brighter a star is, the smaller its magnitude; consequently star 1, the dimmer of the two in this example, has the larger magnitude).

The expression can be rewritten in terms of the brightness ratio:

$$b_2/b_1 = 10^{(m_1 - m_2)/2.5} = 10^{0.4(m_1 - m_2)}$$

In sum, if the magnitude difference between two stars is known, we can use this expression to find their brightness ratio. For example, suppose star 1 has a magnitude of 14.33 and star 2 has a magnitude of 11.86. Then the term in the exponent becomes $0.4(14.33 - 11.86)$, which equals 0.988, and the brightness ratio is $b_2/b_1 = 10^{0.988} = 9.73$.

---

2. Several devices can disperse light according to wavelength; the resulting spectrum may be recorded on film or an electronic detector
3. Stars are classified according to the appearance of their spectra
    a. Early spectroscopists found that the patterns of dark lines in stellar spectra varied from star to star; in the 1860s, Italian astronomer Father Angelo Secchi found that most stars fell into a small number of groups or categories having different patterns of dark lines
    b. In the 1890s, a group of astronomers at Harvard University began a systematic study of the line patterns in stellar spectra; among the several women working in this group, the most prominent was Annie J. Cannon, who is credited with creating the modern spectral classification system
    c. At first, astronomers thought that the differing patterns of spectral lines might mean that stars had different chemical compositions
        (1) Thus, stars having strong lines of hydrogen, the first (and simplest) element, were designated "A" stars, while those with strong lines of atomic helium (the second element) were "B" stars, and so on
        (2) In many cases, the prominent lines were not yet associated with specific elements, and arbitrary letters were assigned

d. Cannon found that, if different types of spectra are arranged in a certain order, patterns of spectral lines change smoothly from one to the next
      (1) Using the letters that already had been assigned to the different types of spectra, the order of this sequence was O, B, A, F, G, K, M
      (2) Cannon was able to recognize fine gradations in the pattern of lines from one class to the next, and developed such subclasses as A1, A2, and so on, for stars that fell between the major types
      (3) Cannon's system, with minor modifications, is still used today, although the understanding of the significance of the sequence has changed
4. Spectral classes are actually groupings of stars according to surface temperature
   a. The relationship between the appearance of the spectrum and temperature is due to ionization; most stars have the same composition
      (1) The hotter a star, the more highly ionized the gas in its outer layers
      (2) The degree of ionization determines which spectral lines are formed
   b. The sequence O, B, A, F, G, K, M is a sequence according to temperature, from hottest (O) to coolest (M)
      (1) O stars, with surface temperatures greater than 30,000 K, are characterized by spectral lines of ionized helium, while M stars, as cool as 2,000 K, have strong spectral lines as a result of having molecules in their atmospheres
      (2) The Sun is a G2 star, intermediate in temperature (about 6,000 K)
   c. Some stars do not fall into any of the standard spectral classes
      (1) These generally are stars in short-lived phases of their lives, which accounts for their rarity
      (2) Some of these unusual stars are characterized by emission lines (instead of the normal absorption-line spectrum), which means that they are surrounded by clouds of hot, thin gas
      (3) Some have unusual chemical compositions
      (4) Some are **variable stars;** these are stars that physically pulsate, alternately expanding and contracting
      (5) Some are losing or acquiring material from a close companion in a double-star system

E. Binary systems
1. More than half of all stars belong to systems of two or more stars; double-star systems are called **binary stars**
   a. In a true binary, the stars are gravitationally bound to each other and orbit a common center of mass
   b. Stars that appear close together in the sky but are not in mutual orbit are called *optical doubles;* these are not true binaries
2. In a **visual binary,** both stars can be seen separately in a telescope
   a. To be detected, the pair must have a relatively wide separation
   b. Only a small fraction of all binaries are visual binaries
      (1) The second star usually is too faint or too close to its brighter companion to be detected
      (2) In these cases, the presence of the binary companion is inferred from other observational clues
3. An **astrometric binary** is identified when the known star undergoes detectable orbital motion

a. Detection of an astrometric binary requires precise positional measurements over a period of time, so that several orbital cycles are observed
b. Because every star has its own random motion across the sky, the path of a star in an astrometric binary is a combination of orbital motion and straight-line motion of the center of mass, resulting in a wavy pattern of motion

4. An *eclipsing binary* is recognized by fluctuations in its brightness as the two stars alternately pass in front of each other, blocking some of the light from the eclipsed star (see *Eclipsing Binary*, page 106)
    a. An eclipsing binary system must have its orbital plane aligned with our line of sight
    b. Eclipsing binaries are rare because such a perfect alignment occurs in only a small fraction of all binary systems

5. A *spectrum binary* is a system in which the two stars have comparable brightness but different *spectral types*, so that a composite spectrum is seen
    a. It is physically impossible for a single star to have spectral lines as a result of different ionization states
    b. These systems also are rare because stars normally having different spectral types also have very different luminosities, and one star blocks the light of the other

6. A *spectroscopic binary* shows Doppler shifts in the spectral lines, thereby revealing orbital motion
    a. Alternating blueshifts and redshifts of the spectral lines reveal orbital motion, as the stars alternately approach and recede along the line of sight
    b. Spectroscopic binaries are by far the most common because Doppler shifts can be measured in any star whose spectrum can be observed
        (1) Only binary systems we see head-on would not be recognized as spectroscopic binaries, because no Doppler shift exists when the motion is perpendicular to the line of sight
        (2) Most binary systems are so far away, or have the two stars so close together, that it is impossible to measure the motion of the visible star directly; however, the Doppler shift can still be detected
    c. In a *single-lined spectroscopic binary*, the presence of a binary companion is inferred from spectral lines undergoing periodic Doppler shifts as the known star orbits its unseen companion
    d. In a *double-lined spectroscopic binary*, spectral lines from both stars are seen because the two stars have comparable brightness; the lines alternately split and merge as the two stars orbit

II. Fundamental Stellar Properties

A. General information
1. Derivation of stellar properties is based on observations, using methods described in the first part of this chapter
2. Scientists do not consider distance to be a basic physical property of a star, but in most cases distance must be calculated before deriving any basic physical information

B. Absolute magnitude and luminosity
1. The *luminosity* of a star is the total energy it emits per second

Eclipsing Binary

Eclipsing binary systems are observed when our line of sight happens to be aligned with the orbital plane of a double-star system; the two stars will alternately eclipse each other. This results in brightness variations in the total light output from the system. In the upper diagram, the smaller star is moving in front of the larger one; in the lower diagram, the smaller one is passing behind. The segmented graph in each case (called the light curve) shows how the total brightness of the one system varies as the eclipses occur.

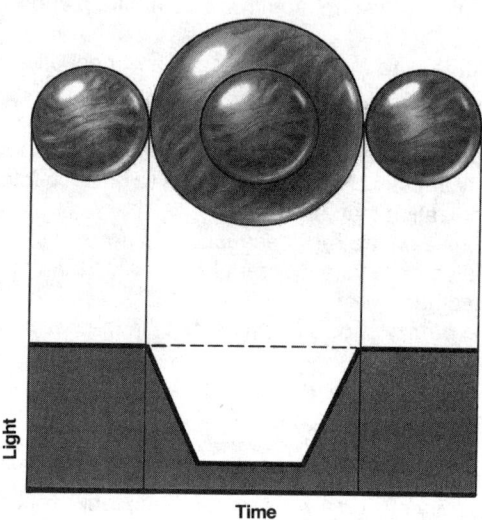

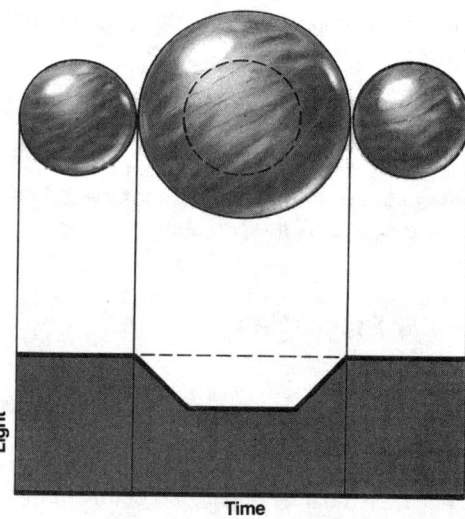

a. Deriving the luminosity of a star requires observations over the entire range of wavelengths at which the star emits, including ultraviolet and infrared ones
   b. Measuring luminosity also requires knowledge of a star's distance so that the apparent brightness can be used to determine the energy emitted at the source
2. Astronomers commonly use the magnitude system as a means of expressing luminosity by defining a standard magnitude that does not depend on the distance
   a. The **absolute magnitude** (M) of a star is the magnitude it would have if it were 10 pc away
   b. The **apparent magnitude** (m) is the observed magnitude of the star at its actual distance
   c. Thus, the absolute magnitude of a star is related to its luminosity; the use of absolute magnitudes allows luminosities of stars to be compared with one another, independently of distance
   d. To determine the absolute magnitude of a star, scientists use the apparent magnitude and distance, employing the inverse square law to determine what the magnitude of the star would be if it were 10 pc away
      (1) The inverse square law states that the apparent brightness of an object must be inversely proportional to the square of its distance ($b = 1/d^2$)
      (2) For example: if a star has apparent magnitude of 12 and is 1,000 pc away, it would be $(100)^2 = 10,000$ times brighter if it were 10 pc away (a factor of 100 closer); because a factor of 10,000 corresponds to 10 magnitudes (there is a factor of 100 in brightness for each 5 magnitudes), the star would be 10 magnitudes brighter at 10 pc than it is at 1,000 pc; thus, its absolute magnitude (M) is $12 - 10 = 2$
      (3) Another example: if a star has apparent magnitude of 4 and is 1 pc away, it would be 10^2 (100) times fainter at a distance of 10 pc; a factor of 100 corresponds to 5 magnitudes, so its absolute magnitude would be 5 magnitudes fainter than 4; thus, its absolute magnitude (M) is $4 + 5 = 9$
   e. Once the absolute magnitude of a star is known, then its luminosity is estimated by comparing it with other stars whose luminosities have been measured and whose absolute magnitudes are known
      (1) For example: it is known that a star having an absolute magnitude of 5 has a luminosity of 4×10^{26} W (these figures represent the Sun); therefore, a star having absolute magnitude of 3 is two magnitudes, or a factor of $(2.5)^2 = 6.25$ times, more luminous; hence, this star has luminosity (L) of $= (6.25) \times (4 \times 10^{26}) = 25 \times 10^{26}$ W
      (2) Another example: a star with M equals 9 is 4 magnitudes less luminous than the star with a magnitude of 5, so it is $(2.5)^4 = 38$ times less luminous, and its luminosity is $(4 \times 10^{26})/(38) = 1.52 \times 10^{28}$ W
      (3) Generally, astronomers express luminosities in units of the Sun's luminosity; the Sun has $M = 5$ and $L = 4 \times 10^{26}$ W, so the stars in the two preceding examples have luminosities of 15.6 and 0.0026 solar luminosities, respectively
   f. The luminosity is the stellar property that has the greatest range of variation from star to star

(1) The most luminous stars have luminosities greater than 1 million Suns, while the least luminous stars are 1/1,000 of the Sun's luminosity
(2) Thus, the full range of luminosity from least to most is a factor of approximately 1 billion

C. The Hertzsprung-Russell diagram
1. In the early 20th century, astronomers Ejnar Hertzsprung and Henry Norris Russell independently discovered the relationship between a star's surface temperature and its luminosity
2. Russell first constructed a diagram showing the relationship between luminosity (expressed in terms of absolute magnitude) and surface temperature (expressed in terms of spectral type); this is called the **Hertzsprung-Russell (H-R) diagram**
   a. An H-R diagram may use any measure of surface temperature on the horizontal axis, with hot stars on the left and cooler ones on the right (see *A Modern H-R Diagram*)
   b. Luminosity commonly is used in place of absolute magnitude on the vertical scale, with a higher luminosity at the top and a lower luminosity at the bottom
3. Stars fall into well-defined positions on the H-R diagram instead of being randomly distributed, meaning that a definite relationship exists between luminosity and surface temperature
   a. The vast majority of all stars lie along a diagonal band from the upper left to the lower right in the H-R diagram; this is called the **main sequence**
   b. A few stars lie above the main sequence, having higher luminosity for their temperature
      (1) These stars have similar surface temperatures to main-sequence stars and therefore must emit the same energy per square meter of surface area; because they are more luminous than main-sequence stars, they must have greater surface area
      (2) Thus, the stars in the upper right-hand area of the H-R diagram are called **giant stars** or **supergiant stars,** because they have greater surface area and larger radii than the stars on the main sequence
   c. A very small number of stars are found in the lower left-hand corner of the H-R diagram, meaning that they have lower luminosity for their surface temperature than main-sequence stars
      (1) These stars must therefore be smaller than main-sequence stars
      (2) Because they also are relatively hot, they are called **white dwarfs**
   d. Stars can be classified according to **luminosity class** because of their different luminosities (in addition to classification by spectral type)
      (1) Class I refers to the most luminous supergiants; class II, bright giants; Class III, giants; class IV, subgiant stars; and class V, main-sequence stars
      (2) The luminosity class of a star often can be inferred from details about the appearance of its spectral lines
      (3) Thus, the full spectral classification of a star includes the spectral type and the luminosity class; in this system, the Sun is a G2V star, meaning that it is a G2 star lying on the main sequence
4. The H-R diagram can be used to find distances to stars

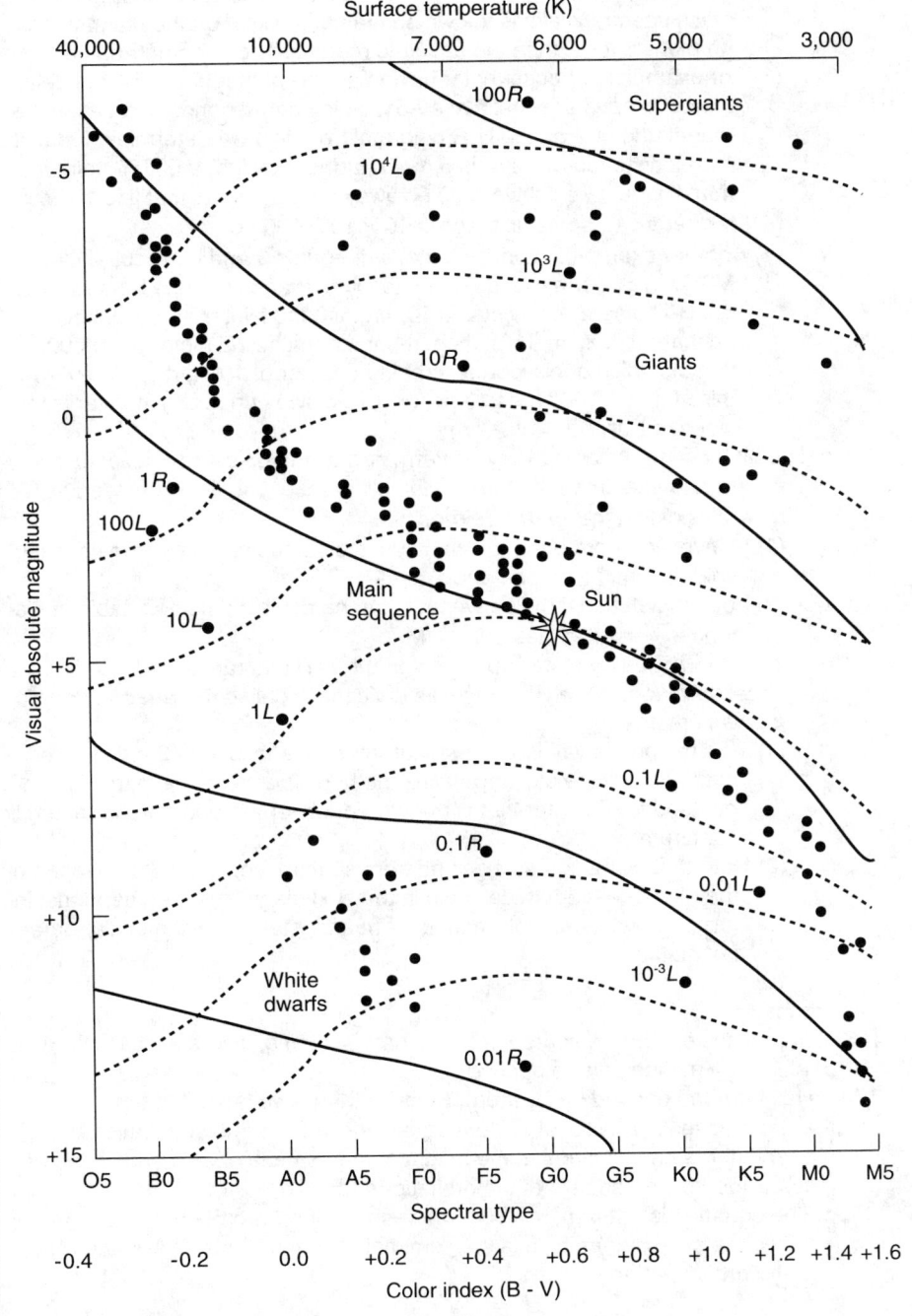

A Modern H-R Diagram

Three commonly used temperature indicators on the horizontal axis are shown in this model H-R diagram. Solid lines show stellar radii in solar units while dotted lines indicate luminosities, also shown in solar units.

a. The distance to a star can be found if its luminosity is known because the apparent brightness depends on how far away it is
b. Astronomers typically compare the absolute and apparent magnitudes of a star to find its distance
   (1) The apparent magnitude can always be measured using a telescope
   (2) The absolute magnitude can be derived from the H-R diagram if the star's spectral class is known so that its position on the diagram can be plotted; if so, then the absolute magnitude can be determined
   (3) For example: suppose a star having apparent magnitude of 7 is determined to have spectral class A5V, which corresponds to an absolute magnitude of 2 on the H-R diagram; from this we determine that at its actual distance this star is 5 magnitudes, or a factor of 100, fainter than it would be if it were 10 pc away, which means that it is 10 times farther away, so its distance is $10 \times 10 = 100$ pc
   (4) Another example: suppose a star's m equals 1 and its spectral class is M3V, which corresponds to an absolute magnitude of 11, as seen in the H-R diagram; this star is 10 magnitudes fainter at a distance of 10 pc than at its actual distance, meaning that it is a factor of 10,000 fainter, which corresponds to being a factor of 100 farther away, so this star is 1/100 times 10 pc, or 0.1 pc away (in reality, no star is this close to Earth)
c. The distance to a star is always determined by the difference between its apparent and absolute magnitudes; this difference, usually expressed as m – M, is called the **distance modulus**
   (1) Conversion from distance modulus to distance can be achieved using calculations described above
   (2) For simple cases, the conversion can be achieved using a table of distances versus values of (m – M)
d. The method of using a star's position on the H-R diagram in order to find its absolute magnitude and, hence, its distance is called the **spectroscopic parallax** method
   (1) This method is valid for stars at greater distances than the distances derived from the stellar parallax method; spectroscopic parallax can be used for any star that is bright enough for astronomers to record its spectrum
   (2) Note that distances determined from spectroscopic parallax depend on the absolute magnitude scale in the H-R diagram, which depends, in turn, on distances determined for nearby stars from the stellar parallax method

D. Stellar masses
1. Stellar masses can be measured only in binary systems, where Kepler's third law of planetary motion can be applied
2. In its full form (as derived by Newton), Kepler's third law states that the sum of the two masses in an orbital system equals the cube of the semimajor axis of the orbit (if the orbit is circular, the length of the semimajor axis equals the radius) divided by the square of the orbital period
   a. The equation is: $m_1 + m_2 = a^3/P^2$, where m_1 and m_2 are the masses of the two stars in solar units, a is the semimajor axis in terms of AU, and P is the orbital period in years

Fundamental Stellar Properties

      b. The period and the semimajor axis are observable quantities
          (1) Only the angular size of the orbit can be measured directly; to convert from this to a true size requires knowing the distance
          (2) Another difficulty in measuring the semimajor axis is that the orbital plane of the binary system affects the apparent angular size of the orbit, making the orientation difficult to determine
          (3) The best-determined cases are eclipsing, double-lined spectroscopic binaries, where the orientation and the orbital velocities (demonstrated by the Doppler shift) are known
   3. If both a and P can be measured, then the sum of the masses can be found from Kepler's third law
      a. In order to find the individual masses of the two stars, more information is needed
      b. If the ratio of the two masses can be determined along with the sum of the masses, then the individual masses can be found; the ratio can be derived only if the relative distances of the two stars from the center of mass can be measured, which is possible only in a visual astrometric binary
      c. In many cases, the only way to estimate individual masses is to assume that one of the stars has the same mass as other stars of its spectral class; this is particularly useful in systems where only one of the two stars is visible and the mass of the unseen companion is sought
   4. The range of masses found for stars is relatively narrow, compared to other stellar properties
      a. The masses range from about 0.08 to 50 times the mass of the Sun
      b. Thus, the full range is a factor of less than 100 from smallest to largest

E. Stellar radii
   1. The most widely used method for measuring stellar radii involves measurements of eclipsing binary systems
      a. The basic method is to see how long it takes for one star to pass in front of the other, and to combine this time with the known orbital velocity to compute the diameter of the star being eclipsed (using the formula distance = speed × time)
      b. The time required for one star to pass in front of the other is simply equal to the duration of the observed eclipse and is easy to measure because the binary system appears to dim during this time
      c. The orbital speed is measured through the use of the Doppler effect, as any eclipsing binary will also be a spectroscopic binary
   2. Another direct method for measuring stellar radii is interferometry
      a. This method utilizes two or more telescopes to observe the same star, allowing light waves entering the separate telescopes to interfere with each other, creating a pattern that depends on the angular size of the star
      b. This method works only for relatively bright stars
      c. In order to convert the observed angular diameter into a true diameter, the distance to the star must be known
          (1) If the distance (d) is measured in parsecs, and the angular diameter of the star is in arcseconds, then the diameter (D) in meters is given by $D = (1.5 \times 10^{11})(d)$

(2) Thus, for example, if a star 100 pc away has an angular diameter of 0.005 arcsecond, then its diameter is $(1.5 \times 10^{11})(0.005)(100) = 7.5 \times 10^{10}$ m (or about 107 times the Sun's diameter)
3. The values found for stellar radii cover a relatively narrow range
   a. Along the main sequence of the H-R diagram, stars vary in radius only from about 0.3 to about 20 times the Sun's radius
   b. Supergiants can be as large as nearly 1,000 solar radii
   c. White dwarfs are usually about 0.01 solar radii in size; that is, they are about the size of Earth

F. Stellar temperatures
1. Measurement of the surface temperatures of stars involves the comparison of a star's brightness at different wavelengths and makes use of Wien's law (see Chapter 4, Light and the Atom)
2. When measuring magnitudes, filters can be used to separate wavelength regions in the spectra of stars
   a. One commonly used filter allows only light in the visible range, from about 500 nm to 600 nm, to enter the photometer
   b. Called the *V magnitude* or visual magnitude, this visible spectrum measurement corresponds closely to what the human eye can see
   c. Another commonly used filter allows light from 400 to 500 nm to enter the photometer; the resulting magnitude is called the blue magnitude, or *B magnitude*
   d. A hot star will be brighter in the blue portion of the spectrum than in the visual portion, whereas a cooler star will be brighter in the visual portion than in the blue portion
      (1) Therefore, the difference between the B and V magnitudes, called the **color index,** is a measure of stellar temperature
      (2) A hot star will have (B – V) less than zero because a brighter B magnitude means that B is smaller than V
      (3) A cooler star will have (B – V) greater than zero because a fainter B magnitude means that B is greater than V
3. The spectral type also can be used to estimate stellar temperature
   a. Spectral type is related to ionization, which in turn depends on temperature
   b. This provides only a crude estimate of surface temperature
4. The most accurate values for stellar surface temperature are derived from the analysis of ionization balance
   a. This analysis requires the identification of many spectral lines
   b. It also requires a detailed computation to determine the relative quantities of different ions from the relative strengths of the spectral lines
5. The range of stellar surface temperatures is quite narrow; the coolest stars have temperatures around 2,000 to 3,000 K, while the hottest (except for the unusual, short-lived phases of some stars) are normally no higher than 40,000 K

G. Stellar composition
1. The relative abundances of the elements affect the strengths of the absorption lines formed by the outer layers of gas in a stellar atmosphere, so spectral line strength is useful in determining chemical composition
   a. In general, the stronger the line, the more abundant the element that forms the line

      b. However, line strength also is affected by ionization and by the intrinsic absorbing strength (or photon absorption probability) of the atom or ion doing the absorbing, which can vary a great deal from one element to another
2. Computed simulations of stellar spectra obtain accurate results, in which the observed line strengths are matched with calculations
      a. These calculations require large, fast computers, and normally are carried out only for selected stars
      b. In the process of calculating a model spectrum, scientists derive other parameters, such as surface density and temperature; the result is a comprehensive picture of the star's outer layers
3. The results of stellar composition measurements show that most stars have essentially the same composition as the Sun
      a. Roughly 73% of the mass of a typical star is composed of hydrogen; 25% is helium; and the remaining 2% is made up of all the other elements
      b. This composition is therefore often referred to as the *cosmic abundances* of the elements

H. Other stellar properties
1. The rotation velocity of a star can be measured using the Doppler effect
      a. Spectral lines are broadened by stellar rotation, because one side of the star is rotating toward us at the same time the other side is rotating away; thus, part of the line is blueshifted while another part is redshifted
      b. The unknown orientation of the rotation axis of a star makes these measurements ambiguous because they cannot measure motion that is perpendicular to the line of sight
2. The surface magnetic field strength of a star can be estimated from the splitting of spectral lines
      a. Certain atoms and ions have electron energy levels that are split in the presence of a magnetic field
      b. The degree of line splitting is proportional to the strength of the magnetic field, so the field strength can be inferred from measurements of line splitting

Study Activities

1. Summarize the applications of the three types of stellar observations.
2. Explain how positional measurements are used to determine distances to stars and to detect binary systems.
3. Determine how much fainter (by what numerical factor) a star of magnitude 9 is compared to a star of magnitude 3, and how much brighter a star of magnitude 2 is compared to a star of magnitude 18.
4. Explain why measurement of most of the basic stellar parameters requires knowledge of stellar distances.
5. Explain the significance of the fact that stars fall into well-defined areas on the H-R diagram instead of being randomly distributed.
6. Use the Stefan-Boltzmann law to explain why stars lying above the main sequence must be larger than main-sequence stars, and stars lying below the main sequence must be smaller than main-sequence stars.

13

Stellar Structure and Evolution

Objectives

After studying this chapter, the reader should be able to:
- Explain how astronomers study the internal structure of stars.
- Explain why the mass of a star is its most fundamental property.
- Summarize the internal structure of stars with different masses.
- Explain how stars generate energy in their cores.
- Explain how the steps in stellar evolution can be deduced from observations.
- Summarize the evolution of a star like the Sun from formation to final state.
- Compare the evolution of more massive stars with the chronology of a star like the Sun.
- Explain how the evolution of a star is altered if it belongs to a close binary system where mass exchange occurs.

I. Stellar Structure

A. General information
1. The internal structure of stars is determined using the same techniques discussed for studying the internal structure of the Sun (see Chapter 11, The Sun)
2. The Sun provides a good basis for comparison, and a great deal is learned about stars through the extension of knowledge about the Sun
3. Stellar structure is investigated using a combination of observations and theory

B. The role of stellar mass
1. Calculations show that the mass of a star is the single most important quantity that determines all the other properties of the star
   a. The luminosity, temperature, and radius are determined by the mass, provided the chemical composition is the same
   b. The statement that the mass and the chemical composition determine all the other basic properties of a star is known as the *Russell-Vogt theorem*
2. The main sequence of the Hertzsprung-Russell (H-R) diagram can be more accurately described as a "mass sequence," with the most massive stars at the upper left in the H-R diagram and the least massive ones at the lower right of the diagram
   a. This is a manifestation of the *mass-luminosity relationship,* a correlation between stellar mass and luminosity
   b. Stars off of the main sequence (giants, supergiants, and white dwarfs) do not obey the same mass-luminosity relationship as the main-sequence stars, primarily because they have different chemical compositions in their cores

3. The dependence of other stellar properties on mass can be understood in terms of ***hydrostatic equilibrium,*** which is the balance that must occur at every point inside a star between the inward force of gravity and the outward force of pressure
   a. If this balance did not occur at some point inside a star, the star would either expand or contract; consequently, a star that is stable must be in balance
   b. Using the concept of hydrostatic equilibrium, one can explain in simple terms why the mass of a star governs its other properties
      (1) The mass of a star determines the strength of the inward force of gravity and the counterbalancing force of pressure
      (2) Pressure is linked to density and temperature, so the mass governs these two properties in the star's interior
      (3) The nuclear reaction rate is governed by the temperature in the core region of the star
      (4) The luminosity of the star is equal to the energy produced in the core's nuclear reactions
      (5) The radius of the star is determined by the distribution of the mass inside it; that is, by the variation of density from the center to the surface

C. Results of model calculations
1. A full stellar model calculation invokes not only hydrostatic equilibrium, but several other relationships describing physical processes that occur inside a star
2. Inside a star, the density, pressure, and temperature all increase dramatically toward the center
3. The chemical composition of all stars is similar
   a. Hydrogen is the most abundant element, followed by helium; all other elements make up only a small fraction of the total
   b. The composition of the core changes with time as a result of nuclear reactions
4. Some processes vary from star to star, depending on the mass of the star
   a. The specific reaction steps in the conversion of hydrogen to helium in the core depend on stellar mass
      (1) For stars on the lower part of the main sequence (just above the Sun's position), the dominant reaction is the ***proton-proton chain,*** which involves only hydrogen nuclei (see Chapter 11, The Sun)
      (2) For stars on the upper portion of the main sequence, the ***CNO cycle*** (or carbon-nitrogen-oxygen cycle) is the dominant reaction; this involves carbon as a catalyst and produces both nitrogen and oxygen during the cycle (see *The CNO Cycle,* page 116)
      (3) For both reactions, the net result is the same: four hydrogen nuclei (protons) fuse to form one helium nucleus (two protons and two neutrons), with a fraction (0.007) of the original mass converted to energy according to the equation $E = mc^2$
      (4) Other reactions occur in stars not on the main sequence; these are described later
   b. Energy transport also varies with stellar mass
      (1) Whether convection or radiative transport dominates depends on how rapidly temperature changes with depth inside a star; in places where it changes rapidly, convection normally dominates

The CNO Cycle

Although the CNO cycle has the same net result as the proton-proton chain, it has several steps and occurs only in massive stars; the net result is that four hydrogen nuclei (protons, designated as $^{1}_{1}H$) are fused together to form one helium nucleus ($^{4}_{2}He$) and energy. Although carbon is required in the sequence, it is neither consumed nor converted into anything else if the process goes to completion; consequently, carbon is a catalyst in the reaction.

Here are the steps of the CNO cycle:

$^{1}_{1}H + {}^{12}_{6}C \rightarrow {}^{13}_{7}N + \gamma$

$^{13}_{7}N \rightarrow {}^{13}_{6}C + e^{+} + \nu$

$^{13}_{6}C + {}^{1}_{1}H \rightarrow {}^{14}_{7}N + \gamma$

$^{14}_{7}N + {}^{1}_{1}H \rightarrow {}^{15}_{8}O + \gamma$

$^{15}_{8}O \rightarrow {}^{15}_{7}N + e^{+} + \nu$

$^{15}_{7}N + {}^{1}_{1}H \rightarrow {}^{12}_{6}C + {}^{4}_{2}He$

This reaction produces more energy than the proton-proton chain at high temperatures; in more massive upper-portion main-sequence stars with greater gravitational heating in their cores, most of the energy comes from the CNO cycle.

---

    (2) For stars on the lower portion of the main sequence, including the Sun, radiative transport dominates in the core while convection takes place in the outer layers

    (3) For upper-portion main-sequence stars, convective transport occurs in the core region while radiative transport dominates in the outer layers

5. Like the Sun, other stars have activity in and above their photospheres
   a. Stars on the lower part of the main sequence have chromospheres and coronae because they have convection in their outer layers, which creates heating above the surface
   b. Stars on the upper portion of the main sequence have high-velocity stellar winds
      (1) These winds are much faster and denser than solar wind, having speeds up to 2,000 or 3,000 km/second and carrying away significant quantities of mass
      (2) The mass loss, which can be as high as the equivalent of one solar mass every 100,000 years, can affect the evolution of a star
      (3) The cause of the winds is not fully understood, but it is known that their acceleration to high speeds is caused by **radiation pressure** — the force exerted as photons of radiation are absorbed by atoms and ions in a star's outer atmosphere
   c. Giants and supergiants in the upper right-hand region of the H-R diagram have chromospheres and strong stellar winds
      (1) These winds have a lower speed (about 20 km/sec) but much higher density than the winds in luminous hot stars
      (2) The mass loss from red giants and supergiants can also be very significant, affecting the evolution of these stars

D. Nuclear reactions and the generation of energy

1. Most of a star's lifetime is spent on the main sequence, when the conversion of hydrogen to helium is the primary nuclear process in the core
   a. The fact that roughly 90% of all stars are on the main sequence indicates that about 90% of a star's lifetime is spent there
   b. The exact steps in the conversion of hydrogen to helium depend on a star's mass, which is determined by its position on the main sequence
2. The lifetime of a star is essentially equal to the length of time it spends on the main sequence
   a. The lifetime of a star is proportional to the total energy available from the conversion of hydrogen to helium in its core, which is proportional to its mass, and inversely proportional to the rate of energy loss, which is equal to its luminosity
   b. The lifetime of a star can be derived by multiplying the factor (M/L) by the Sun's lifetime of 1×10^{10} (10 billion) years, where M and L are the mass and luminosity of the star in solar units, respectively
      (1) For example, an upper-portion main-sequence star having an M of 20 solar masses and an L of 10,000 will have a hydrogen-burning lifetime of $(20/10{,}000) \times (1 \times 10^{10}) = 2 \times 10^7$ years
      (2) Another example: a star on the lower part of the main sequence having an M of 0.1 and an L of 0.001 has a hydrogen-burning lifetime of $(0.1/0.001) \times (1 \times 10^{10}) = 10^{12}$ years (note that this is far longer than the current age of the universe)
      (3) The range of main-sequence stellar lifetimes is from several hundred billion years for low-mass stars to less than 1 million years for the most massive stars; this is one reason that massive stars are rare
3. Other reactions can occur after a star converts all of its core hydrogen into helium
   a. The star undergoes major changes when this occurs, as described later in this chapter
   b. Following the conversion of hydrogen to helium, helium can react to form carbon through the *triple-alpha reaction*
      (1) In this reaction, three helium nuclei, often called *alpha particles*, having two protons and two neutrons each, combine to form a carbon nucleus, containing six protons and six neutrons; a fraction of the original mass is converted into energy according to the equation $E = mc^2$ (see *Triple-Alpha Reaction,* page 118)
      (2) This reaction requires a higher temperature than the previous stage, because the helium nuclei have greater electrical charges than single protons, and must therefore collide at even higher speeds in order to overcome their electrical repulsion and react with each other
   c. After some of the helium has been converted into carbon, a series of reactions called *alpha capture reactions* can occur
      (1) In these reactions, alpha particles are captured by nuclei, building up heavier and heavier nuclei
      (2) For example, when a carbon nucleus captures an alpha particle, the result is an oxygen nucleus, containing 8 protons and 8 neutrons; subsequent captures produce neon, magnesium, and so on
      (3) Each reaction requires a higher temperature, and each produces energy through the conversion of mass according to the equation $E = mc^2$

> **Triple-Alpha Reaction**
>
> The triple-alpha reaction consists of two steps, but they must follow in very rapid succession. Consequently, it is almost correct to think of them as a single reaction involving three particles. In the first step, two helium nuclei (^{4}He) fuse to form an unstable beryllium nucleus (^{8}Be), which normally breaks apart to form two helium nuclei very quickly. Therefore, the third helium nucleus must collide with the ^{8}Be nucleus immediately after it forms.
>
> Here are the steps:
>
> $^4_2$He + $^4_2$He → $^8_4$Be
>
> $^8_4$Be + $^4_2$He → $^{12}_{6}$C + γ.

   d. Other kinds of reactions may begin to occur if the core temperature in the star becomes high enough
      (1) One class of reaction, which can produce very heavy elements, is the **neutron capture reaction,** in which free neutrons are captured, thereby creating heavier nuclei
      (2) Once elements have built up to the point where the stellar core is made of iron (26 protons and 30 neutrons), further reactions no longer produce excess energy, and this has catastrophic consequences, such as supernova explosions

II. Observations of Stellar Evolution

A. General information
   1. The physical structure of a star changes along with its composition; this causes the star to evolve
      a. In terms of hydrostatic equilibrium, the core density increases as lightweight nuclei merge to form heavier ones
      b. As the nuclei in the core of the star merge, the inward force of gravity can compress the core further
   2. Some of the steps in stellar evolution occur rapidly enough so that changes can be observed
   3. Scientists can deduce the sequence of changes in a star's life by observing stars that are in different stages; theoretical stellar structure calculations help determine stellar evolution

B. Star clusters
   1. Clusters of stars provide especially useful opportunities to observe the effects of evolution on stars in different stages of their lives
   2. A star cluster is a grouping of a few to many thousands of stars that are gravitationally bound to each other, and therefore orbit a common center of mass
   3. General types of clusters are found in our galaxy
      a. A galactic, or **open cluster,** is a loose conglomeration of up to hundreds of stars, generally found in the plane of the disk of the galaxy
      b. A **globular cluster** is an enormous spherical ball of stars containing hundreds of thousands of members, typically found above or below the plane of the galaxy's disk, in a region called the galactic **halo**

c. An ***OB association*** is a loosely bound (or possibly unbound) grouping containing very massive, hot, young stars (spectral types O and B)
4. Scientist make certain assumptions about star clusters that are helpful in studying stellar evolution
   a. They assume that stars formed together and have the same age
   b. Stars also formed from the same material and therefore had the same chemical composition initially
   c. Because the size of a star cluster is always much smaller than the distance to the cluster, in effect it can be assumed that all the stars are at the same distance from Earth
5. It is helpful to plot an H-R diagram for a cluster of stars (see *H-R Diagram for Star Clusters,* page 120)
   a. Often the color index is used instead of spectral type to indicate stellar surface temperature (color index is defined in Chapter 12, Measuring the Stars; it refers to the difference between the B, or blue, magnitude, and the V, or visual, magnitude)
   b. Instead of absolute magnitude, apparent magnitude is used to indicate luminosity; since all the stars are at the same distance, their relative luminosities are reflected by their relative apparent magnitudes
   c. A cluster H-R diagram in which apparent magnitude V is plotted against color index (B − V) is called a ***color-magnitude diagram***
   d. In such a diagram, the upper part of the main sequence is often missing, while stars are seen in the upper right-hand region
      (1) From this it is deduced that stars become red giants and supergiants after completing their main-sequence lifetimes
      (2) The most massive stars (those at the upper end of the main sequence) complete their main-sequence lifetimes and become red giants or supergiants most rapidly, while less massive main-sequence stars take longer
      (3) The age of a cluster of stars can be estimated by noting how far down the *main sequence turn-off* lies; this is the point at which the main sequence terminates because all stars above that point have had time to complete the hydrogen-burning phase in their cores
      (4) The older the cluster, the farther down and to the right the main sequence turn-off lies in the color-magnitude diagram
      (5) Thus, to estimate a cluster's age and to learn how stars of a given age have evolved, scientists measure the apparent magnitudes in the visual (V) and blue (B) wavelength bands, and plot a color-magnitude diagram
   e. A cluster color-magnitude diagram can be used to find the distance to the cluster
      (1) This technique, called *main sequence fitting,* involves determining the difference between apparent and absolute magnitudes for the cluster diagram
      (2) To do this, one assumes that the cluster's main sequence matches the H-R diagram's main sequence, so in effect the cluster diagram is superimposed on the standard H-R diagram and slid vertically until the main sequences match; then one can read the difference between apparent magnitude (m) and absolute magnitude (M) from the vertical axis

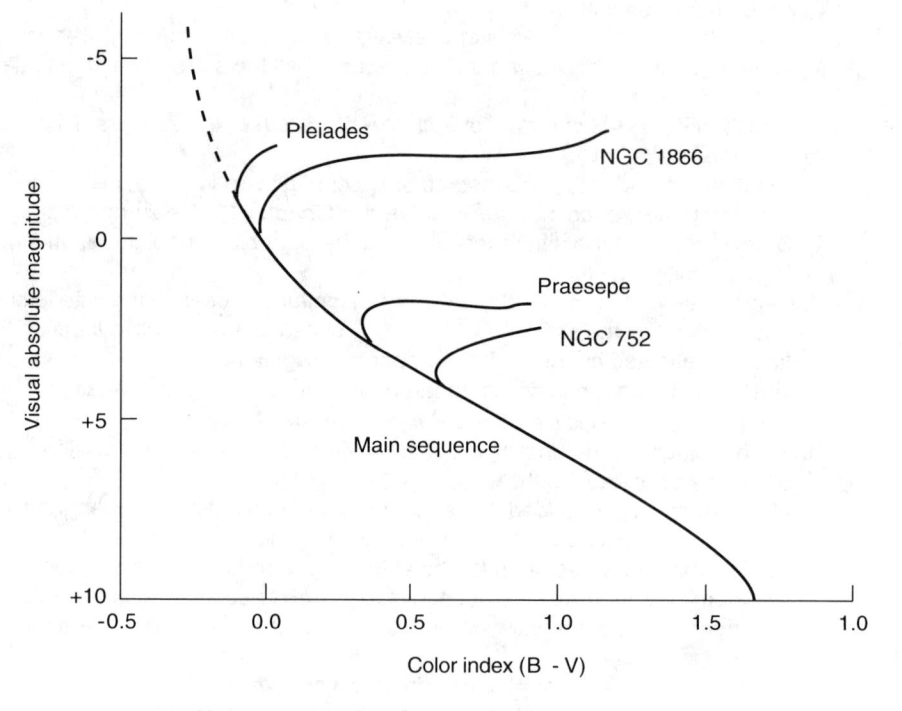

H-R Diagram for Star Clusters

Several star clusters are plotted on the same axes of H-R diagrams (using color index rather than spectral type). Different symbols in the right-hand portion distinguish among the giant stars in different clusters. Main sequence turn-off points are shown as solid lines, which are related to the ages of the clusters.

(3) This difference (m − M) is the distance modulus, from which the cluster distance can be determined, as described in Chapter 12, Measuring the Stars

C. Star formation
1. Infrared observations can provide a great deal of information about the formation of stars
    a. Infrared wavelengths penetrate through dense interstellar clouds much more easily than visible light, and star formation takes place inside dense clouds
    b. Newly formed pre-stellar objects have surface temperatures of a few hundred K and therefore emit most of their energy in the infrared wavelength (recall Wien's law)
    c. Infrared telescopes detect numerous young stars inside dark interstellar clouds, thus demonstrating that stars form from interstellar material
2. Star formation has been developed from a combination of observation and theory
    a. The process begins with the gravitational collapse of a dense interstellar cloud

b. The collapse may begin spontaneously if a random fluctuation due to cloud turbulence causes the density in some region to be high enough that gravity overcomes the cloud pressure
c. The collapse also may be triggered by compression caused by an outside force, such as a shock wave from a stellar explosion
d. Once the collapse begins, it proceeds much more quickly in the central part of the cloud than in the outer regions, and a dense central core quickly forms
e. At first, the core stays cold because infrared emission carries away all the heat energy gained by the gravitational compression
f. Once the core surpasses a certain density, the infrared radiation is trapped, and the core heats up while the contraction slows
g. A second rapid collapse phase may occur if molecular hydrogen (H_2) becomes hot enough to break apart into individual hydrogen atoms; then, heat is no longer trapped, and collapse follows
h. Eventually, the central core becomes so dense that gas pressure nearly counteracts gravity, and the core once again enters a phase of very slow contraction and heating
i. The *protostar,* as it is now called, finally reaches hydrostatic equilibrium and becomes a true star when its core becomes hot enough for nuclear reactions to begin

III. Life Stories of Stars

A. General information
1. A star, such as the Sun, spends most of its lifetime, or about 10 billion years, as a main-sequence star
2. During this time the star produces energy by fusing hydrogen into helium via the proton-proton chain
3. As hydrogen changes to helium, the composition of the core gradually changes, and this affects the structure of the star

B. Evolution of stars like the Sun
1. As the stellar core becomes more compressed and hotter, the higher temperature increases the nuclear reaction rate, thus increasing the star's luminosity
2. As the luminosity gradually increases, the star rises a little on the H-R diagram; for this reason the main sequence is a broad strip, rather than a narrow line on the H-R diagram
3. The lower left edge of the main sequence, where stars lie when nuclear reactions first begin, is called the *zero-age main sequence*
4. When the core hydrogen is all used up, the star becomes a red giant, undergoing dramatic changes in structure (see *Evolutionary Track of a Star Like the Sun,* page 122)
    a. As gravity continues to compress the inner regions of the star, a shell outside the core becomes hot enough to initiate the proton-proton chain reaction, and hydrogen burning begins in that shell
    b. The production of energy from a shell outside the core causes the outer layers of the star to expand

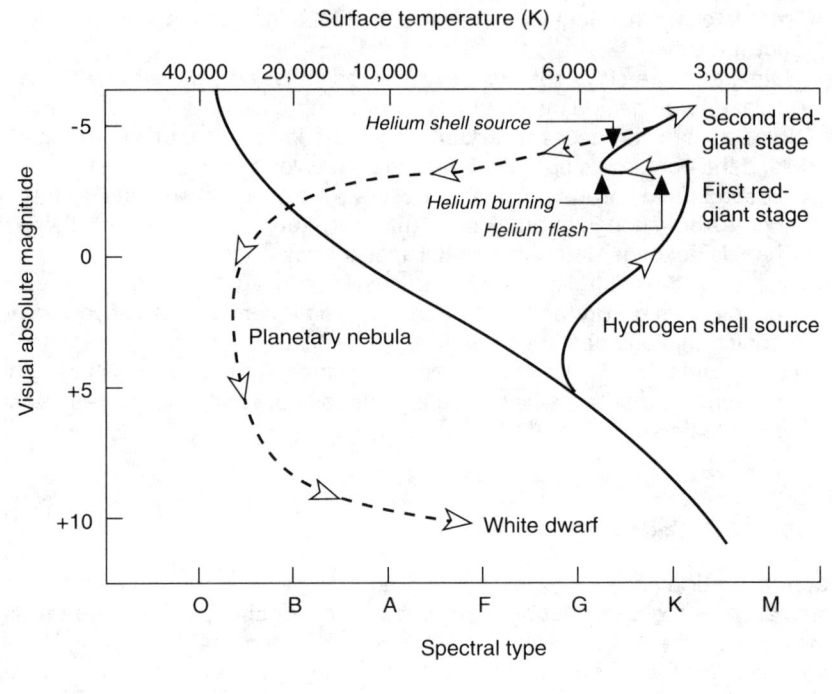

Evolutionary Track of a Star Like the Sun

The lifetime of a star with one solar mass is shown in this H-R diagram. Following the second red-giant stage is the dashed portion; astronomers do not know as much about the dashed portion as they do about the earlier stages.

    (1) The energy outside the core cannot be efficiently transported to the surface and is trapped, thus causing the expansion
    (2) As the expansion proceeds, the outermost layers cool
    (3) The star increases in luminosity while decreasing in surface temperature; it moves up and to the right on the H-R diagram, into the red-giant region
  c. The core stops reacting and continues to shrink and get hotter
  d. As the core shrinks, the free electrons there form a **degenerate electron gas,** which has unusual properties
    (1) In a degenerate gas, the individual particles are prevented from coming any closer together by a principle of quantum mechanics that states that no two particles can occupy the same energy state
    (2) This resistance creates a pressure that now supports the core of the star against further contraction
    (3) Unlike ordinary gas pressure, degenerate electron gas pressure is not affected by temperature changes; thus, the core does not expand in response to any further heat input, as ordinary gas would
5. The core remains inert until it gets hot enough for the triple-alpha reaction to begin converting helium into carbon

       a. The onset of this reaction is sudden and violent because of the degenerate nature of the stellar core
- (1) The core heats further when the new reactions begin to occur; degenerate electron gas cannot expand and cool to counteract this heating
- (2) The increased temperature, in turn, increases the reaction rate, thus causing further heating
- (3) A runaway reaction occurs throughout the entire stellar core in a matter of seconds; this is called the *helium flash*
- (4) The helium flash is not observable because it takes place deep inside the core
- (5) The extra heat from the reactions soon destroys the degeneracy of the core (by creating new energy states for the electrons to occupy), and the core reaches a new equilibrium in which energy is produced by stable helium-burning reactions

       b. As the triple-alpha reaction stabilizes in the core, the outer layers of the star contract and become hotter, and the star moves down and to the left on the H-R diagram

       c. Reactions in the hydrogen-burning shell outside the core are halted at this time because the helium flash causes expansion and cooling of the shell

6. The star reaches its final state as the helium in the core is used up
   a. A helium shell-burning source ignites, causing a second red-giant phase as the outer layers expand
   b. During this second red-giant phase the star loses mass from its outer layers via stellar wind
      - (1) The wind can occur steadily and gradually, depleting the outer layers of the star
      - (2) As the outer layers are removed and hot core is exposed, a rapid, low-density wind moves outward and compresses the slower-moving material from the red-giant wind into a thin shell, forming a **planetary nebula**
      - (3) Planetary nebulae commonly are observed, and often are very striking in appearance, because the shell of thin gas surrounding the central star glows via emission lines (recall Kirchhoff's rules), producing a red (as a result of hydrogen) or blue-green (as a result of ionized oxygen) color
   c. When the helium in the core is used up, the core again contracts and becomes degenerate; the outer layers of the star are ejected
   d. The star moves to the lower left-hand corner of the H-R diagram as its outer layers are lost

7. In this final state, the star is a **white dwarf;** this is the exposed core of a star in which all nuclear reactions have stopped
   a. A white dwarf contains approximately the same mass as the Sun, but it is as small as Earth
      - (1) The density of a white dwarf is immense; it is roughly 1×10^6 g/cm^3 — about a million times the density of water
      - (2) The surface temperature of a white dwarf is high, but the size is so small that it has very little surface area and the luminosity is low; hence, it lies in the lower left-hand corner of the H-R diagram
   b. A white dwarf typically undergoes no further changes, except for gradually cooling off and becoming dimmer

C. Evolution of stars that have 2 to 8 solar masses

1. As in the case of a star that has 1 solar mass, the more massive star spends most of its lifetime as a main-sequence star (converting hydrogen into helium in its core)
   a. The more massive star will lie higher up on the main sequence than the star with 1 solar mass; this kind of star ranges from type F (for 2 solar masses) to type B (for 8 solar masses)
   b. In this case, the dominant hydrogen-burning reaction is the CNO cycle (not the proton-proton chain)
   c. The lifetime of a more massive star on the main sequence is shorter than that of a star with 1 solar mass
2. When the hydrogen in the core is used up, the more massive star will become a red giant
   a. Hydrogen burning begins in a shell outside the core, causing the outer layers to expand
   b. Because the core temperature is higher than in the star with 1 solar mass, the core does not become degenerate, and there is no helium flash when the triple-alpha reaction begins in the core
   c. As helium burning proceeds, the star's outer layers contract and the star moves to the left on the H-R diagram
3. The more massive star will undergo additional reaction stages in its core
   a. Because of the higher mass of this star (as compared to a star with 1 solar mass), its core temperature and density will be high enough to initiate new reaction stages after some of the helium is used up
   b. The new reactions, which begin while the helium is still burning, are alpha capture reactions, and they convert the core first to oxygen, then neon, then magnesium, and so on
4. The star undergoes a new red-giant phase each time a nuclear fuel is exhausted and a new shell source ignites outside of the core; thus, the star loops back and forth in the upper right-hand corner of the H-R diagram
5. As a red giant, the star is losing mass via stellar wind; this loss of mass counteracts the tendency of the star's core to keep getting hotter and more compressed
6. Finally, the star will become a white dwarf whose composition is determined by the number of reaction stages that were completed
   a. For stars with initial solar masses between 2 and 6, the end result is a carbon-oxygen white dwarf
   b. For stars with initial solar masses between 6 and 8, the end result is an oxygen-neon-magnesium white dwarf
   c. However, there is a limit on the highest mass a white dwarf can achieve because degenerate electron gas pressure can support only so much mass; the *Chandrasekhar limit* for white dwarf mass is 1.44 solar masses

D. Evolution of the more massive stars

1. The phases in the evolution of a very massive star occur much more quickly than for stars of lower mass
2. A massive star spends most of its lifetime on the main sequence, while it converts hydrogen into helium in its core

3. A massive star undergoes many nuclear-burning stages as a result of the high temperatures that are reached in the core (caused by gravitational compression)
   a. Thus, these stars undergo multiple red-giant phases
   b. Ultimately, iron is produced in the core of a massive star
4. Once the core has been converted into iron, further reaction stages require more energy than they produce and the star undergoes catastrophic changes
   a. As these reactions begin, the stellar core collapses violently
   b. The outcome of the core collapse depends on how much mass it contains
5. The iron core collapses, leading to a **supernova** explosion if the core mass is no more than approximately 3 solar masses
   a. As collapse occurs, protons and electrons are forced to come together, thus forming neutrons
      (1) The process in which protons and electrons are forced to come together is called an inverse beta decay
      (2) In addition to the neutron that is the principal end product of inverse beta decay, a neutrino also is released
   b. Neutrons, like electrons, cannot occupy identical energy states, and this creates a **degenerate neutron gas** whose pressure is sufficient to halt the collapse of the core
   c. The core is now called a **neutron star,** having a mass of 2 or 3 solar masses but compressed into a sphere having a diameter of only about 10 km
   d. The neutron star is extremely rigid; when the outer layers of the original star crash down onto its surface, they rebound with so much energy that the outer portion of the star explodes
   e. During the supernova explosion, very rapid nuclear reactions take place, adding new elements to those already being dispersed into interstellar space
6. If the iron core contains more than 3 solar masses, its collapse leads to the formation of a **black hole**
   a. In this case, the degenerate neutron gas pressure cannot halt the collapse
   b. There may or may not be a supernova explosion, depending on whether a neutron star forms temporarily (causing a rebound of the infalling outer layers of the star) before collapsing further
   c. A black hole never stops collapsing; mathematically, it can be described as a single point containing all the mass of the collapsed stellar core, but physically it is difficult to describe (its properties are described in Chapter 14, Stellar Remnants)

E. The effects of mass transfer in binary systems
1. In binary star systems with stars that are close together, matter can be transferred from one star to the other
2. The transfer of mass affects the evolution of both stars
   a. The star that loses mass has its evolution slowed because there is less internal compression and heating
   b. The star that gains mass has its evolution speeded up and it will undergo more nuclear reactions
3. The result of mass transfer can be a binary system in which the less massive star has reached a later stage in evolution than the more massive star

a. This contradicts normal expectations because generally the more massive star evolves more quickly
   b. The explanation is that the star that is now less massive began as the one that was more massive, and it evolved to the red-giant phase before losing some of its mass
4. Binary systems exist in which the mass transfer goes first from one star to the other, and then is reversed when the second star becomes a red giant and develops a stellar wind of its own

Study Activities

1. Explain why the study of the Sun's structure and evolution is useful in investigating the evolution of stars of other masses.
2. Explain why we can be certain that a star is in hydrostatic equilibrium at every point in its interior.
3. Calculate the hydrogen-burning lifetimes for the following:
   (a) a star having 5 times the Sun's mass and 500 times its luminosity
   (b) a star having one-half the Sun's mass and 0.03 times its luminosity.
4. Summarize the contrasts between a star on the upper portion of the main sequence of the H-R diagram and one on the lower portion of the main sequence in the following areas: nuclear reactions, energy transport, lifetime on the main sequence, and surface activity (in terms of chromospheres, coronae, and stellar winds).

14

Stellar Remnants

Objectives

After studying this chapter, the reader should be able to:
- Describe the basic properties of a white dwarf.
- Explain how white dwarfs can be observed.
- Describe the properties of supernova remnants.
- Summarize the properties of a neutron star.
- Explain how neutron stars are observed.
- Describe the formation and observation of accretion disks in mass-transfer binary systems.
- Summarize the properties of a black hole.
- Explain how black holes are sought observationally, and summarize the current evidence of their existence.

I. White Dwarfs

A. General information
1. White dwarfs are the remnants of stars whose original masses range up to about 8 times the mass of the Sun
2. Because of their small size, white dwarfs have low luminosities and are difficult to observe directly
3. The first white dwarf ever detected is Sirius B, the companion to Sirius, the brightest star in the sky
   a. It was known that Sirius had a companion because the orbital motion of Sirius was measured (it is an astrometric binary)
   b. When careful observations finally revealed the dim companion, astronomers realized the star must be very small in order to be so dim, and the term white dwarf was introduced

B. Properties of white dwarfs
1. A white dwarf is supported by degenerate electron gas pressure creating hydrostatic equilibrium
2. With a mass near 1 solar mass and a radius comparable to that of Earth, a white dwarf has an extremely high density
   a. Typically, the density is about 1×10^6 g/cm^3, or about one million times the density of water

      b. It is impossible to create matter in this form on Earth; therefore, it can be studied only through theoretical considerations and observations of white dwarfs
  3. Because it is the exposed core of a star, a white dwarf initially is very hot
      a. Typical surface temperatures for white dwarfs are 10,000 K or higher
      b. The temperature is uniform throughout the interior because the electrons there are free to move about, carrying energy with them in a process called conduction
  4. The composition of a white dwarf is determined by the last nuclear reaction stage the star underwent before becoming a white dwarf (see Chapter 12, Measuring the Stars); it may become a carbon-oxygen-helium white dwarf or an oxygen-neon-magnesium white dwarf
  5. A white dwarf has a thin layer of ordinary gas surrounding the degenerate interior
      a. This gas may include remains of the original outer layers of the star and may represent the original hydrogen-rich composition
      b. An alternative theory is that the gas is the remnant of a nuclear reaction that occurred in a shell outside the core of the original star
      c. Because of the high surface gravity of a white dwarf, heavy elements can quickly sink below the surface, leaving a surface layer that is stratified according to elemental weight, with the lightest element at the top
  6. The thin layer of ordinary gas surrounding a white dwarf traps heat inside the star
      a. This gas can transport energy only by the radiative process, which is very inefficient
         (1) Heat transport by radiation is slow
         (2) The surface area of a white dwarf is small, so energy also is radiated away into space at a low rate
      b. Calculations show that it can take billions of years to cool down to the point where it no longer glows at visible wavelengths

C. Observations of isolated white dwarfs

  1. Nearby white dwarfs are bright enough to be observed directly, but only with large telescopes in most cases
      a. A large fraction of the stars in the vicinity of the Sun are white dwarfs
      b. This suggests that a significant fraction of the mass of the entire galaxy is in this form
  2. The spectral lines in white dwarfs are shifted toward long wavelengths and are broadened
      a. The shift toward long wavelengths (as compared to laboratory measurements of the same spectral lines) is due to the **gravitational redshift**
         (1) The gravitational redshift was predicted by Einstein's theory of general relativity
         (2) When escaping from a very strong field, such as that of a white dwarf, photons lose energy and are thus shifted to longer wavelengths
      b. The reason that spectral lines in white dwarfs are broad is that the pressure in the atmosphere is very high as a result of the strong surface gravity
         (1) The close packing and frequent collisions of the particles distort the electron energy levels, so that electrons in a given energy level can have a range of energies
         (2) Thus, the absorption lines formed by these electrons are widened; this process is called *pressure broadening*

       c. Another reason that spectral lines in white dwarfs are broad is that these stars sometimes have very strong magnetic fields
          (1) The magnetic fields can be millions of times stronger than the Sun's magnetic field
          (2) These strong fields form as a result of contraction of the original star as it becomes a white dwarf; the original magnetic field is retained, but squeezed into a smaller volume, thus increasing its intensity
          (3) The magnetic field causes the *Zeeman effect,* in which certain spectral lines are split in the presence of a magnetic field, with the breadth of the splitting proportional to the strength of the field
    3. Once a white dwarf has cooled off and is no longer visible, it is called a *black dwarf*
       a. A black dwarf cannot be observed directly, except possibly as an infrared source
       b. It is difficult to determine how many black dwarfs exist; some astronomers estimate that they may be very common

D. Observations of white dwarfs in mass-transfer binary systems
    1. If a white dwarf acquires additional material, it may experience a sudden increase in brightness that is readily observable
       a. The acquisition of new material can occur in a binary system where a white dwarf receives matter from its companion star
       b. In much more rare circumstances, a white dwarf may acquire new material through a collision with a cloud of interstellar gas
    2. Degenerate gas causes a white dwarf to flare up when new material is added
       a. If a degenerate gas is heated, it does not respond by expanding and cooling (see Chapter 13, Stellar Structure and Evolution)
       b. Thus, if new matter is added and allows nuclear reactions to begin, there is a rapid runaway effect, and the white dwarf suddenly flares up in brightness
    3. The detailed nature of the outburst depends on how much matter is added
       a. The quantity of matter acquired by a white dwarf in a binary system depends on the details of the mass-transfer process
       b. The mass transfer normally occurs through the formation of an **accretion disk**
          (1) An accretion disk is formed as matter coming from the companion star orbits the white dwarf
          (2) Material from the accretion disk spirals onto the white dwarf as a result of friction and collisions among gas particles in the disk
          (3) The accretion disk may develop instabilities that allow material to build up for a time, and then fall onto the white dwarf gradually or in clumps
    4. If a large quantity of matter is added to a white dwarf at once, a major nuclear reaction occurs at the surface, creating a ***nova***
       a. At its peak, a nova may reach a luminosity approaching 1×10^5 times the luminosity of the Sun
       b. The same white dwarf may become a nova repeatedly, if new material is added on separate occasions
          (1) Most novae take hundreds or thousands of years to repeat the cycle of mass accretion and surface explosion
          (2) Some especially massive white dwarfs flare up on timescales of a few decades and are called *recurrent novae*

5. If a small quantity of matter is added, more minor, nonnuclear flare-ups occur, creating *dwarf novae* or one of several types of *cataclysmic variable*
   a. The outbursts in such systems are due to released gravitational potential energy, not nuclear reactions, and are far less luminous than those in novae
   b. These outbursts may be separated by only days or hours in a given system
6. If a carbon white dwarf receives enough new material to exceed the mass limit for white dwarfs, the entire star explodes in a supernova
   a. A supernova created by an exploding carbon-oxygen white dwarf is called a **Type I supernova**
      (1) Very rapid nuclear reactions in a Type I supernova create heavy elements, which are then dispersed into space
      (2) Most of the iron in the galaxy is the product of Type I supernovae
   b. An entirely different supernova process—one in which a massive star develops an iron core which then collapses and rebounds—is called a **Type II supernova**

II. Supernova Remnants

A. General information
   1. When a supernova (of either type) occurs, an expanding cloud of gas, called a *supernova remnant,* is left behind
   2. Many supernova remnants are observed throughout the galaxy
      a. The principal means of detecting them is through radio observations
      b. The number of observed supernova remnants, combined with estimates of how long they persist, indicates that supernovae occur in the galaxy, on average, once every 30 to 50 years
   3. The best known supernova remnant is the Crab Nebula
      a. This turbulent, hot gas cloud coincides with the position of a supernova outburst observed in the year 1054 A.D.
      b. The Crab Nebula is a young supernova remnant, glowing at many wavelengths throughout the electromagnetic spectrum
      c. In the central portion of the Crab Nebula is a neutron star; its discovery helped demonstrate the link between supernova explosions and the formation of neutron stars (described in Chapter 13, Stellar Structure and Evolution)

B. Observations of supernova remnants
   1. Supernova remnants emit **synchrotron radiation** rather than thermal radiation
      a. Synchrotron radiation is created by electrons moving at very high speeds in a magnetic field
      b. Synchrotron radiation can be distinguished from thermal radiation by the fact that it does not have a clearly defined peak emission wavelength and is polarized (see Chapter 4, Light and the Atom)
   2. The presence of synchrotron radiation always means that very high energies are involved because the electrons must be moving near the speed of light
   3. Synchrotron radiation can occur throughout the electromagnetic spectrum; after considerable time, a supernova remnant emits most strongly in the radio portion of the spectrum

a. Thus, only relatively young supernova remnants, such as the Crab Nebula, are observed in X-ray or visible wavelengths
   b. The majority of supernova remnants are known only through radio observations

C. **Role of supernova remnants in the galaxy**
   1. Supernova remnants play an important role in the energy balance of the interstellar medium in the galaxy
      a. A supernova outburst creates powerful shock waves that propagate through the interstellar medium
      b. These shock waves cause heating of the interstellar gas
         (1) The shocks create gas temperatures as hot as 1 million K
         (2) This superheated gas expands and fills much of the volume of interstellar space in the galaxy
      c. The shock waves also cause rapid motions in the interstellar medium; clouds are observed to move about with speeds as high as 100 km/sec
   2. Supernova shock waves also may trigger the formation of new stars
      a. Star formation can be initiated when a shock wave strikes an interstellar cloud, compressing it so that it collapses under its own gravity
      b. This can lead to sequential star formation in regions of dense interstellar clouds, as massive stars form and quickly explode in supernovae, creating shock waves that trigger the formation of new stars, and so on

D. **Evolution of a supernova remnant**
   1. A remnant changes with time, eventually dispersing
      a. Initially, it is very hot and energetic, emitting X-rays, ultraviolet radiation, and visible light as well as infrared and radio radiation
      b. As it ages, a supernova remnant loses energy because of radiation and collisions with interstellar material, finally fading into obscurity as it merges into the general interstellar medium
   2. The typical lifetime of a supernova remnant is 10,000 years

III. Neutron Stars

A. **General information**
   1. The concept of a star made entirely of neutrons was first explored theoretically over 50 years ago
      a. The idea came about through quantum mechanics calculations, which showed that neutrons, like electrons, could not be squeezed into identical energy states and therefore could form a degenerate gas under conditions of sufficiently high pressure
      b. At that time it was not expected that such objects could ever be observed, even if they did exist, because they would be small and dim
   2. The first hint that neutron stars might be real and could be observed occurred in 1967, with the discovery of pulsars (described in part C. below)
   3. Today, the existence of neutron stars is well established

B. **Properties of neutron stars**
   1. The mass of a neutron star ranges from 1 to about 3 solar masses

a. Less massive stellar remnants typically end up as white dwarfs
b. More massive stellar remnants cannot exist as neutron stars because there is a limit to the pressure that can be exerted by a degenerate neutron gas
c. The exact mass limit depends on how rapidly the neutron star is rotating
2. The radius of a neutron star is approximately 10 km
3. Thus, the density is immense and is far greater than the density of any other material (except an atomic nucleus)
   a. Typical values are in the range of 1×10^{14} to 1×10^{15} g/cm^3; this is more than 100 million times the density of a white dwarf
   b. Such a density is comparable to that of the nucleus of an atom; in a sense, a neutron star can be regarded as a gigantic nucleus
4. The internal structure of a neutron star is governed by the behavior of the degenerate neutron gas
   a. Nearly all of the volume of the star is made of neutrons
   b. Scientists speculate that a very thin (a few centimeters thick) layer of ordinary gas is at the surface
   c. The neutron gas in the interior is thought to form a rigid crystalline structure
5. A neutron star can have a very strong magnetic field resulting from the compression of the star's original field before it collapsed

C. Neutron stars as pulsars
1. In 1967, a new astronomical object called a **pulsar** was discovered
2. A pulsar is characterized by rapid flashes of radiation, primarily at radio wavelengths
   a. The rate of pulsation, or flashing, is very high, with periods typically ranging from less than one-tenth of a second to a few seconds
   b. Some pulsars are now known to radiate (and flash on and off) at X-ray and even gamma-ray wavelengths
3. The rate of flashing is uniform, with very little change
   a. Careful monitoring over years reveals that the rate is slowing very gradually in most cases
   b. Occasionally pulsars undergo "glitches," when the frequency of pulses suddenly increases; this is thought to be caused by a sudden fracture in the crystalline structure
4. It was quickly deduced that only rapidly rotating neutron stars could be responsible for the pulsar phenomenon
   a. The only viable alternative explanation, pulsating white dwarfs, was ruled out because a white dwarf cannot physically expand and contract rapidly enough
   b. Any object larger than a neutron star could not rotate fast enough because its surface velocity would have to exceed the speed of light
   c. Confirmation of the rotating neutron star hypothesis came about through the discovery of a pulsar in the Crab Nebula
      (1) This pulsar was the most rapid, hence the youngest, of those discovered at the time
      (2) As a pulsar slows its spin, it loses energy to its surroundings; the energy being lost by the Crab pulsar was found to match the energy being emitted by the Crab Nebula
5. A pulsar flashes because it emits beams of radiation that sweep across the sky as it spins (see *The Pulsar Mechanism*)

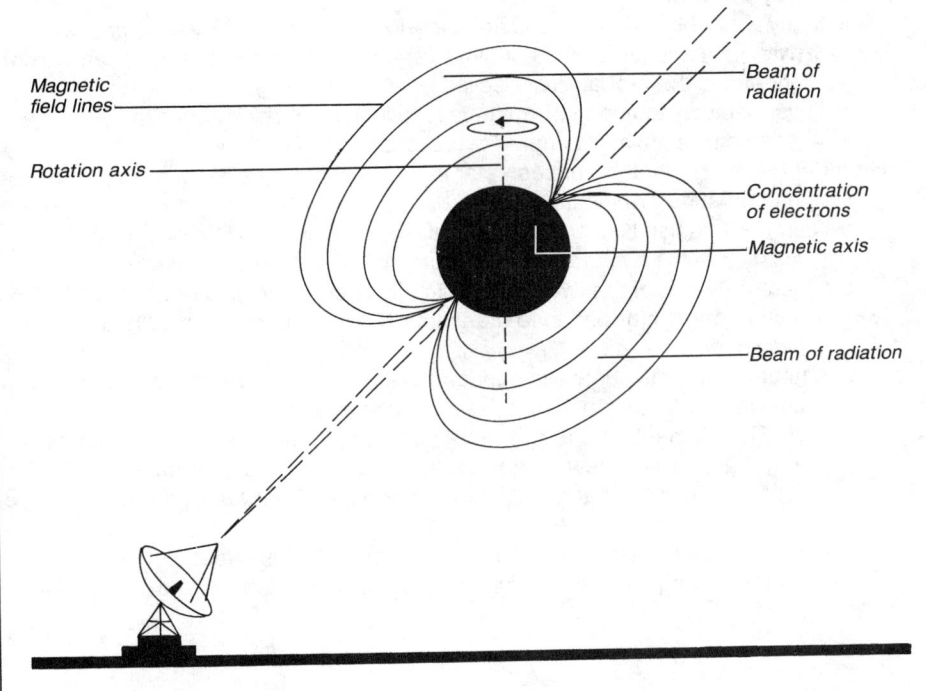

The Pulsar Mechanism

Pulsars are rapidly rotating neutron stars appearing with their magnetic axis out of alignment with the rotational axis. The star gives off synchrotron radiation in narrow beams from above the magnetic poles, where charged particles move along magnetic lines of force. As the star rotates, the beam sweeps the sky; if Earth is positioned in a direction covered by one of the beams, we observe a pulsar.

    a. The beams of radiation are caused by electrons trapped in the neutron star's magnetic field
    b. The trapped electrons travel along the magnetic field lines, which are concentrated at the magnetic poles of the neutron star; hence, the beams are emitted from the magnetic poles
    c. If the magnetic axis is not aligned with the rotation axis of the neutron star, then the beam sweeps across the sky much like a searchlight beam
  6. The beams of radiation sweep across the sky in Earth's direction only if there is a chance alignment; hence, most neutron stars are not seen as pulsars
  7. Scientists have identified a new class of pulsars called millisecond pulsars
    a. *Millisecond pulsars* have periods of a few one-thousandths of a second; they are much more rapid than "normal" pulsars
    b. These pulsars may be old neutron stars in binary systems that have recently acquired additional mass from their companions; the conservation of angular momentum causes them to spin more and more rapidly as they gain mass

D. Neutron stars in mass-transfer binary systems
  1. A neutron star in a close binary system may gain mass from its companion star

2. The mass that is transferred to the neutron star forms an accretion disk around it; because the disk is highly compressed by the star's gravitational field, the disk may be as hot as 100 million K
3. The high temperature of the accretion disk causes it to radiate at X-ray wavelengths
4. The X-ray emission from the accretion disk allows such systems to be observed as **X-ray binaries**
   a. Many X-ray binaries are eclipsing binaries because the tiny neutron star (with its accretion disk) is orbiting very close to a normal, large star, which enhances the probability of eclipse
   b. Thus, one way to find neutron stars is to look for X-ray binaries in which the X-ray emission is interrupted by eclipses
5. In order to confirm that the unseen star in an X-ray binary system is a neutron star, one must determine its mass
   a. Astronomers use Kepler's third law of planetary motion to measure the masses of the two stars (see Chapter 12, Measuring the Stars)
   b. Where only one of the two stars is directly observable, it is not possible to gain enough information to measure precisely the masses of the two stars individually
   c. A neutron star may undergo sudden nuclear reactions on its surface when new matter falls onto it from the accretion disk
      (1) Thus, a neutron star may flare up in a manner similar to a white dwarf that accretes new matter and undergoes a nova outburst
      (2) Neutron stars that occasionally flare up at X-ray wavelengths are called *bursters*
   d. If a neutron star gains enough mass to exceed the limit that can be supported by degenerate neutron gas pressure, it collapses and forms a black hole

IV. Black Holes

A. General information
1. The idea that an object could collapse into a volume so small that its gravitational field would trap light was developed soon after Einstein's theory of general relativity was published in 1915
2. The term "black hole" was coined to describe such an object more than 50 years ago, long before there was any evidence that such objects existed
3. Today, there is ample evidence that black holes exist

B. Properties of black holes
1. If a star having more than 2 or 3 solar masses in its core collapses, it will exceed the mass limit for formation of a neutron star
   a. When a star collapses beyond the point where degenerate neutron gas pressure can support it, the collapse never stops
   b. Thus, a black hole is not in hydrostatic equilibrium because there is no known force that can counteract the inward force of gravity
   c. It is said that the mass of the star forms a *singularity,* described mathematically as a single point having infinite density
2. As the collapse proceeds, the surface gravity of the star becomes stronger

a. The gravitational force of the star remains the same at distances outside of the original surface of the star; the immense increase in gravity occurs only at closer distances
   b. As the surface gravity increases, it has an increasingly significant effect on photons of light
   c. Eventually a point is reached where the surface gravity is so great that light cannot escape
      (1) At this point, the star is said to have passed through the *event horizon* because it is impossible to observe anything that happens to it after this
      (2) The radius of the star at this point is called the *Schwarzschild radius*
      (3) The Schwarzschild radius is proportional to the mass of the star; for a star of 1 solar mass, it is 3 km

C. **Observational evidence for black holes**
   1. A black hole cannot be directly observed, but its presence may be detected through its gravitational effects
   2. If a binary system is found to have an unseen member whose mass is too great to be a neutron star, then it must be a black hole
      a. Such binary systems are most easily recognized if mass transfer takes place from the companion star to the black hole
      b. In this case, the matter that is transferred forms an accretion disk so hot that it emits X-rays
      c. Thus, X-ray binaries are likely places to look for black holes
   3. Several X-ray binaries have been observed in which the analysis of the orbit of the visible star indicates that the unseen companion has too much mass to be a neutron star and must therefore be a black hole

Study Activities

1. Explain why a white dwarf has low luminosity, even though it may have a high surface temperature.
2. Discuss the role of degenerate electron gas in creating a nova outburst.
3. Explain how observations can distinguish between thermal and synchrotron radiation from a source, such as a supernova remnant.
4. Explain how supernova explosions can trigger the formation of new stars.
5. Describe the pulsar mechanism; explain how a neutron star creates the rapid flashes that can be observed.
6. Explain why both white dwarfs and neutron stars have mass limits.
7. Explain how a black hole may be distinguished from a white dwarf or neutron star in a mass-transfer binary system.

15

The Milky Way

Objectives

After studying this chapter, the reader should be able to:
- Describe the overall structure, formation, and evolution of the Milky Way galaxy.
- Explain how stellar distances are measured using pulsating variable stars.
- Explain how the 21-cm emission line of hydrogen is used to map the spiral arms of the galaxy.
- Summarize the method for measuring the mass of the galaxy, and describe the evidence for dark matter.
- Summarize the evidence for a supermassive black hole at the core of the galaxy.
- Describe the general properties of interstellar gas and dust.
- Summarize the properties of Population I and Population II stars, and explain how they help astronomers deduce the history of the galaxy.
- Explain how the spiral arms of the galaxy formed and how they are maintained.

I. Overall Structure of the Milky Way

A. General information
1. The Milky Way appears to us as a hazy band of diffuse light that stretches across the sky
   a. This band of light is an edge-on view of a disk-like galaxy containing approximately 100 billion stars
   b. From our position, embedded within the disk of the galaxy, it is impossible to get an overall view
2. During the first three decades of the 20th century, astronomers gradually deduced the size and structure of the galaxy
   a. Initial counts of the number of stars as a function of distance from Earth made it appear that Earth's location is near the center of the galaxy
      (1) The density of stars appeared to drop off with increasing distance, making it appear that we are located in a central, dense region
      (2) It was later found that this appearance was caused by interstellar dust, which makes stars appear fainter, hence more distant, than they should
   b. Evidence that the solar system is far from the center of the galaxy came from analyses of globular clusters and stellar motions in the 1920s
      (1) Harlow Shapley measured distances to a number of globular clusters and found that they are distributed around a center that is far from the

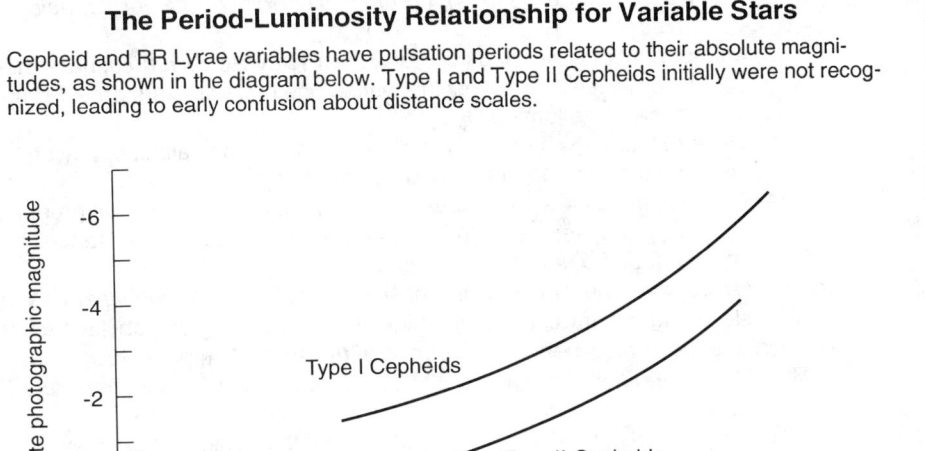

The Period-Luminosity Relationship for Variable Stars

Cepheid and RR Lyrae variables have pulsation periods related to their absolute magnitudes, as shown in the diagram below. Type I and Type II Cepheids initially were not recognized, leading to early confusion about distance scales.

        solar system; it was logical to assume that this was the center of mass of the galaxy

      (2) Jan Oort and Bertil Lindblad independently analyzed the motions of stars near the Sun and found that these stars (along with the Sun) are orbiting a center of motion that is far from the Sun, coinciding in direction with the center of the globular cluster distribution

    3. Today, a great deal of information on the structure of the galaxy is derived from radio and infrared observations

      a. These wavelengths have the advantage that they can penetrate the clouds of interstellar gas and dust that block visible light

      b. Radio observations are especially useful in deducing the properties and distribution of interstellar gas

      c. Infrared observations are useful in mapping the distribution of the interstellar dust

B. The disk, central bulge, and halo of the galaxy

    1. Measurements of the size and shape of the galaxy depend on measuring the distances of pulsating variable stars

      a. These stars physically expand and contract with a regular period of pulsation

      b. A definite relationship between the period of pulsation and the luminosity of the star was found that for certain types of pulsating variable stars around the turn of the 20th century

        (1) For one class of stars, the **Cepheid variables,** the period of pulsation is proportional to the luminosity; this **period-luminosity relationship** allows the luminosity of a star to be deduced if the period of pulsation

is measured (see *The Period-Luminosity Relationship for Variable Stars,* page 138)
   (2) Members of another class of stars, the **RR Lyrae variables,** have the same luminosity, so simply identifying one of these stars provides information on its luminosity
   (3) Once the luminosity is known, the distance can be calculated by observing the apparent brightness of the star
   (4) These stars are very luminous giants and can be observed at great distances; thus, their use allowed astronomers to measure distances throughout the galaxy and beyond
2. As seen in a sideways view by a distant observer, the Milky Way would appear as a thin disk having a central, rounded bulge with a scattering of globular clusters surrounding the center (see *The Structure of the Milky Way,* page 139)
   a. The disk has a diameter of about 30,000 pc (or 30 kiloparsecs, abbreviated 30 kpc)
      (1) The Sun is in the disk, about 8.5 kpc from the center of the galaxy
      (2) The bulk of the interstellar gas and dust is confined to a thin plane within the disk of the galaxy
   b. The central bulge is confined to the innermost kiloparsec of the galaxy and is several hundred parsecs thick
      (1) The density of stars in the central bulge is higher than in the outer portions of the disk
      (2) At the core of the central bulge is the *galactic nucleus,* the point about which the rest of the galaxy rotates
   c. The halo of the galaxy is a spherical volume surrounding the central bulge
      (1) The most prominent objects in the halo are the globular clusters
      (2) Isolated stars also are in the halo, but their number is unknown because they are difficult to detect at their great distances from Earth
3. Dim, lower-mass stars far outnumber the more massive, very luminous stars throughout the galaxy
   a. This is because lower-mass stars form more commonly than higher-mass stars
   b. Another reason is that the lower-mass stars live much longer than the massive stars; at any given time, only those massive stars that have formed recently are still in existence, whereas all the lower-mass stars formed are still on the main sequence
4. The mass of the galactic disk is determined using Kepler's third law
   a. A star orbiting far from the galactic center may be treated as though it were orbiting a single, central object whose mass equals the mass of the entire disk inside the star's position
   b. Thus, Kepler's third law can be applied, where the period (P) is the orbital period of the star, the semimajor axis (a) is the orbital radius of the star (the star's distance from the center of the galaxy), and the two masses are the mass of the star and of the galaxy itself
   c. The period and radius of the Sun's orbit can be used in this calculation, because the Sun is far from the galactic center
      (1) The Sun's orbital period, deduced from its velocity (about 220 km/sec) and its distance from the center (about 8.5 kpc, or 1.8×10^9 AU), is roughly 240 million years; substituting these values for P and a into

The Structure of the Milky Way

This illustration shows the characteristic bulge and disk of the Milky Way, according to modern view.

- Spiral arms
- Nucleus
- Sun
- Direction of rotation
- Globular clusters
- Nucleus
- 15,000 pc
- Plane of disk
- Central bulge
- Sun

    the equation derived from Kepler's third law yields a mass of approximately 1×10^{11} solar masses for the galaxy

  (2) This method neglects any mass that lies beyond the Sun's position in the galaxy; thus, it underestimates the total mass

C. The spiral arms
  1. From our location within the disk, it is difficult to observe and map the spiral arms of the galaxy because interstellar dust is concentrated in the plane of the galactic disk and blocks our view

2. Because very luminous O and B stars are concentrated in bands that are part of the spiral arms, astronomers measure the distance to these stars to obtain information about the spiral arm structure
3. The spiral arm structure has been mapped using the 21-cm emission line of hydrogen atoms in interstellar space
   a. Hydrogen atoms have two very closely spaced electron energy levels; when an electron drops from the upper to the lower of these states, it emits a photon with a wavelength of 21.1 cm, in the radio portion of the spectrum
   b. Because hydrogen is the most abundant element in the galaxy, it is a good indicator of spiral arms
   c. The radiation at the 21-cm wavelength of this spectral line is unaffected by interstellar dust, thus allowing the entire spiral arm structure to be mapped
   d. A map of the spiral structure requires a combination of observations and theoretical analysis because the distances to 21-cm emitting clouds cannot be measured directly
      (1) The 21-cm emission line is normally seen to be split into several closely spaced lines by the Doppler effect (see Chapter 4, Light and the Atom); each of these lines is emitted by gas in a different spiral arm, moving at a different speed relative to Earth
      (2) Using the measured velocities of the separate spiral arms and a mathematical model of the rotation speed of the galaxy (as a function of distance from the center), it is possible to derive a map of the spiral arm structure
4. The Milky Way has a fragmentary spiral structure, with segments of spiral arms instead of the continuous spiral arms seen in some galaxies

D. The halo
1. The halo is roughly spherical with a radius of several thousand parsecs
   a. Thus, the halo envelops the entire disk and central bulge
   b. It is very difficult to determine where the outer boundaries of the halo lie
2. The most prominent objects in the halo are the globular clusters
   a. These large, spherical clusters contain several hundred thousand members
   b. The globular clusters are thought to be the oldest objects in the galaxy
      (1) Their ages are estimated from measurements of the main sequence turn-off
      (2) The globular clusters are between 11 and 15 billion years old, which means that they are nearly as old as the universe itself (see Chapter 17, Cosmology)
3. Observations indicate that a substantial amount of mass resides in the outer galaxy and halo
   a. Instead of decreasing as the distance from the center increases, orbital velocities for stars and gas clouds remain high at great distances from the center of the galaxy
      (1) The orbital speeds would decrease with distance if the mass of the galaxy were concentrated at the center
      (2) It is estimated that as much as 90% of the mass of the galaxy resides in the outer portions
   b. We do not know what form the unseen mass in the outer galaxy takes
      (1) One possibility is that the halo is densely populated with such stars as lower main-sequence stars, but this is ruled out because such stars

are not found in sufficient numbers in the portion of the halo nearest the Sun, where they could be seen
- (2) Another possibility is that the halo is densely populated with stellar remnants, such as white dwarfs, neutron stars, or black holes; this hypothesis is difficult to prove
- (3) An additional possibility is that the unseen mass consists of subatomic particles as yet undetected in laboratory experiments
  c. The unseen matter that dominates the mass of the galaxy is known as **dark matter;** there is evidence that dark matter constitutes most of the mass of the universe (see Chapter 17, Cosmology)

E. The core of the galaxy
1. At the center of the galactic bulge is a tiny, luminous object known as the galactic nucleus; its exact size is unknown
2. Radio observations reveal a group of tiny objects in the region of the nucleus
   a. The collective radio source at the core of the galaxy is known as Sagittarius A (Sgr A)
   b. One of the objects in this group, known as Sagittarius A* (Sgr A*), appears to be the true galactic nucleus, the point about which the entire system rotates
3. Evidence indicates that the nucleus is the site of energetic activity
   a. A powerful source of X-ray emission coincides with the position of Sgr A*
   b. High-speed motions of gas clouds near the nucleus indicate the presence of a large quantity of mass
   c. A spiral arm segment close to the galactic center, known as the expanding arm, is moving away from the center; its motion indicates that some explosive event has occurred at the center
   d. Infrared radiation from Sgr A* has been detected recently
4. Whatever the source of the energy from Sgr A*, it is confined to a very small volume; interferometry indicates that it is smaller than the solar system
5. A leading hypothesis among astronomers is that a supermassive black hole resides at the core of the galaxy
   a. The mass, approximately 1 million solar masses, is estimated from the orbital motions of gas clouds near the nucleus
   b. The alternative explanation—that the object may be a very dense cluster of normal stars—appears to be ruled out by recent infrared observations, which should have revealed individual stars in such a cluster
   c. Such a black hole might have formed early in the history of the galaxy
   d. The activity in the nucleus of the Milky Way may be related to the much more energetic activity observed in the cores of some other galaxies (see Chapter 16, Galaxies)

II. The Interstellar Medium

A. General information
1. Interstellar gas and dust comprise approximately 10% to 15% of the total mass in the Milky Way's disk
2. This material is one component of a galactic recycling process
   a. New stars form from the interstellar material

b. As stars age and die, they return material to the interstellar medium; the returned material is often enriched through nuclear processing
3. Interstellar gas and dust are confined to the central plane of the galactic disk
   a. The thickness of the dust plane is less than 200 pc, whereas the stars in the galactic plane are more than several hundred parsecs thick
   b. Thinner interstellar gas exists in the galactic halo
4. Even though the interstellar medium contains a significant fraction of the mass of the galactic disk, it has an enormous volume and an extremely low density
   a. The average density in the galactic disk is only about 1 particle/cm^3, whereas the density of Earth's atmosphere at sea level is about 2×10^{19} particles/cm^3
   b. Even in the densest, darkest interstellar clouds, the density rarely exceeds 1×10^6 particles/cm^3, which is less than one-trillionth of Earth's atmospheric density and is comparable to the best vacuums attainable in laboratory experiments

B. Interstellar dust
1. Most interstellar dust forms in the outer atmospheres of red-giant and supergiant stars
   a. These stars have slow, dense stellar winds, in which the gas cools and condenses, forming tiny solid particles
      (1) The condensation occurs much like the formation of frost on a cold morning; for every chemical compound there is a combination of temperature and pressure at which it will condense from the gas phase into solid form (a similar process took place during the formation of the planets; see Chapter 6, Solar System Formation and Planetary Science)
      (2) The composition of the resulting dust grains depends on which materials condense into solid form most easily; it also depends on the composition of the stellar wind, which depends in turn on the nuclear reaction history of the star (see Chapter 13, Stellar Structure and Evolution)
   b. These dust grains are then expelled into the general interstellar medium by the stellar wind
   c. Some interstellar dust originates in the shells surrounding planetary nebulae and in the expanding gas from nova outbursts
2. One way to measure the properties of the dust in the general interstellar medium is by studying **interstellar extinction,** the dimming effect of the dust on distant stars
   a. Because the dust permeates the entire interstellar medium, the effect of extinction occurs in all distant stars — regardless of direction — although it is more severe toward stars lying behind dense regions
   b. As noted earlier, one effect is to make stars appear more distant than they actually are; this confused early efforts to map the distribution of stars near the Sun
   c. The extinction is strongest for short wavelengths of light and weakest for long wavelengths
      (1) Thus, extinction makes a distant star appear redder in color because the blue wavelengths are diminished more than the red by the interstellar dust; this effect is called *interstellar reddening*

(2) The extinction is particularly severe at ultraviolet wavelengths, making observations of distant stars or stars behind dense clouds difficult at these wavelengths

(3) The extinction is much weaker at infrared wavelengths, allowing observations of stars lying behind large quantities of dust at these wavelengths

3. Interstellar dust also can be observed via thermal emission, which occurs primarily at infrared wavelengths (see Chapter 4, Light and the Atom)
   a. The temperatures of the dust particles typically range from 10 to 50 K, causing them to emit most strongly at wavelengths in the far-infrared spectrum, which can be observed only from space
   b. Dust grains in heated regions, particularly those areas surrounding stars in the process of formation or those in the stellar winds coming from red giants, can have temperatures of several hundred K, and emit most strongly at near-infrared wavelengths, which can be observed from the ground

4. The general properties of interstellar dust grains can be deduced from a combination of interstellar extinction measurements and observations of the thermal emission
   a. The grains range in size from 1×10^{-9} to 1×10^{-6} m, with the smaller sizes far more numerous
   b. The grains are far less numerous, however, than interstellar gas particles
      (1) There is approximately one dust grain for every trillion gas particles
      (2) The total mass of the dust grains represents about 10% of the total mass in the interstellar medium
   c. The composition of the grains depends on the composition of the stellar wind in which they formed
      (1) Grains originating in the outflows from stars with normal hydrogen-dominated outer atmospheres form grains dominated by oxygen-rich compounds, such as silicates
      (2) Grains formed in stars whose outer layers are rich in carbon (such as red giants which have undergone the triple-alpha reaction) are made of carbon or carbon-based compounds

C. The diffuse interstellar medium

1. Most of the volume of the galaxy is occupied by diffuse material, consisting of low-density gas and dust that is transparent at visible wavelengths
2. The principal method for studying this material is to analyze its effects on the light of stars
   a. The gas in the diffuse interstellar medium creates absorption lines, as atoms and ions in space absorb photons of light at specific wavelengths (see Chapter 4, Light and the Atom)
      (1) Interstellar absorption lines can be distinguished from stellar absorption lines because the interstellar lines are more narrow and represent cold gas — they generally arise from atoms and ions that exist at lower temperatures (up to a few hundred K opposed to several thousand K) than stellar photospheres
      (2) The absorption lines can be analyzed to provide information on the composition and state of the diffuse interstellar medium

      b. The properties of the dust in the diffuse interstellar medium are derived by comparing stars lying behind large amounts of dust with stars of similar spectral type that are nearby and not affected by the dust
         (1) The differences in brightness and color between a distant star and a nearby one of similar spectral type are attributed to the effects of the dust because the basic properties of stars of similar spectral type are thought to be identical (see Chapter 12, Measuring the Stars)
         (2) The extinction wavelength can then be determined; this leads, in turn, to estimates of the sizes of the dust particles
  3. The diffuse interstellar medium displays a wide range of properties
      a. The densities of diffuse interstellar clouds range from 1 to 1,000 particles/cm^3
      b. The temperatures of diffuse interstellar clouds range from about 20 to 150 K
      c. A large portion of the diffuse interstellar medium is very hot, with a high degree of ionization and a low density; these regions have temperatures as high as 1×10^6 K
      d. The high-temperature, low-density regions of the interstellar medium are thought to have been created by shock waves from supernova explosions
         (1) The hot, rarefied regions created by supernova explosions form a network of cavities that permeate the galaxy
         (2) The solar system is located within one of these cavities; thus, the Sun and planets are immersed in a medium of low density and high temperature

D. Bright nebulae

1. **Emission nebulae,** or **H II regions,** are interstellar clouds that are heated to the point where they glow
    a. The source of heat is radiation from hot stars that are embedded within the cloud
    b. The term "H II region" refers to the fact that hydrogen (H) is ionized; astronomers use Roman numerals to indicate the degree of ionization, so that H I is atomic hydrogen and H II is ionized hydrogen
2. Emission nebulae often have a red color because much of the radiation is in the form of the red emission line of hydrogen
3. A *reflection nebula* is a cloud lying just behind a hot star, so that light from the star is reflected in our direction by dust grains in the cloud
    a. The reflected light looks blue because the dust grains reflect short-wavelength radiation more efficiently than long-wavelength (red) radiation
    b. In many cases, emission nebulae (which are red) and reflection nebulae (which are blue) are seen surrounding a hot star or cluster of stars

E. Dark clouds

1. The densest interstellar regions are opaque to visible wavelengths of light and appear on photographs as black regions because no stars are seen through them
    a. These dark clouds, also known as molecular clouds, contain the majority of the mass in the interstellar medium
    b. The interiors of dark clouds cannot be observed at visible or ultraviolet wavelengths but can be probed using infrared and radio telescopes
2. Dark clouds are the regions where stars form in the galaxy
3. The bulk of the gas in dark clouds is in molecular form

       a. As in the rest of the galaxy, hydrogen is the most abundant element; in dark clouds, hydrogen takes the form of H_2
       b. Carbon dioxide (CO_2) is the next most abundant molecular species; others also exist in dense clouds
   4. Molecular species in dense clouds are identified according to their emissions at specific radio wavelengths
       a. Most molecular emission lines are found to occur at microwave wavelengths, which are a few millimeters to a centimeter long
       b. Over 100 molecular species have now been identified in dark clouds through their radio-wavelength emission lines

III. Formation and Evolution of the Galaxy

A. General information
1. The ages of globular clusters, estimated from their main sequence turn-off points, suggest that the galaxy is between 13 and 15 billion years old
2. The galaxy has evolved since its formation because there has been sufficient time for many generations of stars to complete their nuclear-reacting lifetimes
    a. One general result has been a gradual enrichment of the abundance of heavy elements in the galaxy
    b. The most massive stars have played the most important role in galactic heavy-element enrichment despite their rarity
       (1) Only the most massive stars go through enough nuclear reaction phases to form the heaviest elements (see Chapter 13, Stellar Structure and Evolution)
       (2) These stars, having high masses, have short lifetimes; there has been time for many generations of them to form, create new elements, and disperse the elements through stellar winds and supernova explosions
3. Much of what is known about the evolution of our own galaxy is based on observations of other galaxies

B. Stellar populations
1. In the 1940s, scientists observed that the nearby spiral galaxy M31 (the Andromeda galaxy) has distinct subsystems of stars that can be distinguished on the basis of color, spectroscopic characteristics, and location within the galaxy
2. Subsequent studies showed that a similar categorization of stars can be applied to the Milky Way
3. Stars are classified by **stellar population**
    a. Population I stars are found in the disk of the galaxy, particularly in the spiral arms
       (1) These stars have "normal" abundances of heavy elements — their compositions are similar to that of the Sun, which has about 2% of its mass in the form of elements other than hydrogen and helium
       (2) Population I stars include massive stars, which have short lifetimes; thus, Population I includes recently formed young stars
    b. Population II stars are found in the halo and the nuclear bulge of the galaxy
       (1) These stars have low abundances of heavy elements; as little as 0.01% of the mass may be in the form of elements heavier than hydrogen and helium

(2) Some Population II stars have highly elliptical orbits that intersect the galactic plane at large angles; these stars are called *high-velocity stars* because they do not have the same rotation as the galactic disk, and the Sun (which does rotate in the same direction as the disk) approaches or recedes from these stars at high speeds

(3) Population II stars are uniformly old; they are not massive, short-lived stars, and they dominate globular clusters, which are the oldest objects in the galaxy

    c. Two distinct populations do not exist; in reality, there is a range of properties between Population I and Population II

    d. Some Population II stars have traces of heavy elements

(1) This indicates that the material from which they formed must have been enriched by the products of previous generations of stars, because scientist believe that no elements heavier than hydrogen and helium existed before the first stars formed

(2) No stars made only of hydrogen and helium have been found; astronomers refer to this missing group as Population III

(3) The lack of stars with no heavy elements may indicate that the very first stars that formed from pure hydrogen and helium were all massive and had short lifetimes; thus, from early times, the galaxy has had some heavy elements available to be incorporated into new stars

C. The formation and maintenance of spiral arms

1. The formation of spiral arms can be explained by simple fluid mechanics
    a. Almost any disturbance to a rotating, fluid disk will create a spiral structure
        (1) The disturbance could be a gravitational force exerted by a neighboring galaxy
        (2) The disturbance could be the blast wave from a supernova explosion
    b. Without a mechanism to maintain the spiral structure, it will soon dissipate
        (1) The spiral arms will be wound tighter and tighter as the galaxy rotates
        (2) In about the time it takes the galaxy to rotate once, the arms will completely disappear
2. A very successful mechanism for maintaining spiral arms is described by the **spiral density wave** theory
    a. This is the same mechanism thought to be responsible for some of the fine spiral structure in the rings of Saturn
    b. This mechanism is based on the fact that standing waves exist in a rotating, fluid disk
        (1) Standing waves are waves that have fixed positions because the distance between waves is a simple fraction of the size of the oscillating medium; they oscillate in place
        (2) The same disturbance that creates spiral structure also creates waves in the rotating disk; interference quickly diminishes all waves except those that have the right wavelength to become standing waves
        (3) The standing waves in a rotating disk have a spiral shape
    c. Spiral density waves are *compressional waves,* meaning that they consist of alternating dense and rarefied regions
        (1) The spiral arms are denser than the space between them

(2) Thus, as a star orbits the galaxy and passes through spiral arms, it tends to spend more time in the arms because the resulting gravitational attraction slows the star as it passes through

(3) The density enhancements are especially strong for interstellar gas and dust, which are confined to the spiral arms; this explains why star formation and the locations of hot, massive, short-lived stars are concentrated in the spiral arms

d. In the simplest case, a spiral density wave has just two spiral arms

(1) Some galaxies have only two arms

(2) The Milky Way, however, has a very complex structure consisting of many spiral arm segments, which suggests that a different mechanism may be at work

3. Sequential star formation in giant molecular cloud complexes provides another mechanism for creating and maintaining a spiral structure

a. Shock waves from supernova explosions can trigger star formation in interstellar clouds

(1) The shock can compress a cloud enough so that it collapses under its own gravity, leading to star formation

(2) This is most likely to happen where the density is high, as in dark interstellar clouds

b. Once star formation begins in a dark cloud, repeated generations of massive stars can cause sequential star formation to gradually consume the cloud

(1) If the cloud is large, it will be stretched into an arclike shape by galactic rotation

(2) Thus, the combination of sequential star formation and rotation of the galaxy produces an arclike structure having all the characteristics of a spiral arm, including a high concentration of interstellar matter and a high number of young, massive stars

c. This mechanism would produce a galaxy having numerous fragments of spiral arms, rather than a symmetrical spiral structure; this is more likely to be the mechanism at work in the Milky Way

D. The formation and evolution of the Milky Way

1. The distribution and ages of the stellar populations, along with theoretical studies of the motions of stars in a disklike system, allow astronomers to develop a scenario to explain how the galaxy formed

2. The first step in the process is thought to have been the gravitational collapse of an enormous cloud of gas

a. As the collapse began, the first stars formed

(1) These stars were formed as the cloud fell in on itself; therefore, they have highly elliptical orbits that are randomly oriented because the cloud was roughly spherical in shape

(2) These stars define the halo of the galaxy

(3) Globular clusters also formed during the earliest times; these ancient clusters have highly elliptical orbits

(4) Because little or no previous star formation had occurred, all stars in the halo were formed with low abundances of heavy elements

b. As the collapse proceeded, the cloud began to be flattened by rotation and formed a disk
   (1) This flattening primarily affected the interstellar material
   (2) The stars already formed continued to follow their orbits, and their distribution continued to fill the spherical volume of the original cloud
c. The central bulge of the galaxy represents the inner, dense region of the original cloud with only a mild degree of flattening
   (1) Hence, the central bulge is denser than the outer disk
   (2) The stars in the central bulge generally are old and have low abundances of heavy elements
d. After a thin disk was formed, star formation continued in the disk
   (1) Stars formed in the disk were given circular orbits lying in the plane of the disk
   (2) Star formation and enrichment by nuclear processes has continued in the disk, with the result that stars formed there have enhanced abundances of heavy elements
   (3) Because the inner portion of the disk is denser than its outer regions, more star formation and nuclear processes have occurred, with the result that the heavy-element abundances are higher there than in the outer portion of the disk
3. The result of the collapse of the original cloud explains the present-day distribution of Population I and Population II stars
   a. Population II stars are those formed during the collapse, before some nuclear processes had occurred and before the galaxy had formed a disk; thus, these stars today are found in the halo and central bulge of the galaxy, are uniformly very old, and have low abundances of heavy elements
   b. Population I stars are those formed after the disk had formed; thus, these stars have circular orbits lying in the plane of the disk, are associated with interstellar gas and dust, are concentrated in spiral arms, include very young stars and star-forming regions, and have higher abundances of heavy elements
4. Some problems exist in this scenario
   a. As mentioned earlier, it is not known why some of the oldest stars have some traces of heavy elements (because the original cloud should have been composed of only hydrogen and helium)
      (1) This gives rise to the problem of the missing Population III stars
      (2) One possible explanation is that a very early generation of massive stars went through their nuclear evolutions and exploded as supernovae even before the formation of the Population II stars
      (3) Another possible explanation is that some heavy elements were formed during the early expansion of the universe; this contradicts modern cosmological theory (see Chapter 17, Cosmology)
   b. Another problem is that globular clusters apparently are not the same age and were not formed at about the same time
      (1) Globular cluster ages range from 11 to 15 billion years, which is too wide a range to allow all of them to have formed during the early collapse phase of the galaxy
      (2) This suggests that the galaxy may have formed from a merger of several smaller systems of stars, which formed separately over a long period of time

5. Regardless of how the galaxy formed, the formation of spiral arms would have occurred in the same fashion, as described in part C. above
6. The galaxy now appears to be in a steady state
   a. Thus, the recycling of matter between stars and the interstellar medium, with ongoing enrichment of heavy elements, will continue indefinitely
   b. It is expected that the form of the galaxy (halo, disk, and spiral arms) also will not change in the foreseeable future

Study Activities

1. Explain how astronomers know that we live in a disklike galaxy at a location far removed from the center.
2. Summarize the reasons that radio observations are especially useful in determining the spiral structure of the galaxy.
3. Compare the maximum distances that can be measured using stellar parallax, spectroscopic parallax, and the period-luminosity relationship for pulsating variable stars.
4. Explain how the mass of the galaxy is determined, and summarize the existing evidence for large quantities of unseen matter.
5. List the forms that the unseen mass in the halo of the galaxy might take.
6. What is the evidence that a supermassive black hole exists at the nucleus of the galaxy?
7. Explain how the properties of interstellar dust grains are deduced from observations.
8. Summarize the various types of physical conditions that exist in the interstellar medium, from the diffuse regions to the dark clouds.
9. Contrast Population I and Population II stars, and explain how these came about during the formation of the galaxy.
10. Explain why the most massive stars, despite their rarity, are responsible for the chemical evolution of the galaxy. Why is it that the enrichment of heavy elements has been accomplished exclusively by massive stars, which make up only a tiny fraction of all stars?

16

Galaxies

Objectives

After studying this chapter, the reader should be able to:
- Describe the Hubble system for classifying galaxies.
- Explain how the basic properties of galaxies are measured.
- Summarize the properties of groups and clusters of galaxies.
- Describe the overall distribution of matter in the universe.
- Explain how the expansion of the universe was discovered.
- Summarize the use of redshifts as distance indicators.
- Describe the characteristics of galaxies that have active nuclei.
- Summarize the properties of quasars, and explain their role in probing the early universe.
- Describe the current interpretation of quasars as young galaxies.

I. Normal Galaxies

A. General information
1. In the early 20th century, astronomers had difficulty settling the issue of whether other galaxies existed outside the Milky Way
    a. Many diffuse "nebulae" were known, but they were considered possible gas clouds within the Milky Way
    b. Some nebulae were spiral-shaped; these were thought to be newly-forming solar systems within our galaxy
2. In 1920, a famous debate took place between Harlow Shapley, who favored the local hypothesis, and Heber Curtis, who favored the interpretation that the nebulae were galaxies
3. The issue was settled in 1926, when Edwin Hubble reported the discovery of Cepheid variable stars in the Andromeda Nebula, using their period-luminosity relationship to demonstrate that the Andromeda Nebula is too distant to be a part of the Milky Way
4. Today, astronomers recognize that galaxies are the fundamental constituents of the universe

B. Classification of galaxies
1. After having demonstrated that "nebulae" are galaxies, Hubble developed a scheme for classifying galaxies according to shape
2. In Hubble's scheme two general classes of galaxies exist: **elliptical galaxies** and **spiral galaxies**

a. Elliptical galaxies are rounded in appearance, with varying degrees of elongation
   (1) An elliptical galaxy is classified by an E with a number following it; this number is determined by the ratio of the length to the width of the galaxy's image (the range is from E0 to E7)
   (2) If a is the long axis (the length) and b the short axis (the width), the numerical classification is determined using the formula $10(1 - b/a)$
   (3) For example, an elliptical galaxy whose length was twice its width would be an E5 galaxy
   (4) Many very small, dim, elliptical or spheroidal galaxies are found in clusters of galaxies; these are called *dwarf elliptical* or *dwarf spheroidal* galaxies
b. Spiral galaxies are classified according to the tightness of the galaxy's spiral arms and the relative size of the central bulge or nucleus
   (1) A spiral with a large central bulge and tightly wound spiral arms is classified as an Sa galaxy
   (2) A spiral with a small central bulge and open spiral arms is classified as an Sc galaxy
   (3) The Milky Way is intermediate between these two extremes and is thought to be an Sb galaxy (this is difficult to determine precisely because we cannot obtain an external view of our own galaxy)
c. About half of all spiral galaxies have an elongated central bulge; these are called **barred spiral galaxies**
   (1) Barred spiral galaxies are classified as SBa, SBb, and SBc, according to the tightness of the spiral arms and the relative size of the central bulge, in analogy with the classification of spiral galaxies
   (2) Recently, it has been determined that the Milky Way has an elongated central bulge, though perhaps not sufficiently elongated to classify it as a barred spiral galaxy
3. A substantial minority (perhaps 15%) of all galaxies do not fall into the elliptical or the spiral classification
   a. Some are disk-shaped galaxies having no spiral structure; these are classified as S0 galaxies
   b. Some show hints of spiral structure but have a disorganized overall appearance; these are called Irregular Type I galaxies
   c. Other galaxies simply do not fit into any of the standard classes and are called Irregular Type II galaxies
      (1) Some Irregular Type II galaxies appear to be deformed as a result of colliding with neighbor galaxies
      (2) Other Irregular Type II galaxies are heavily shrouded in interstellar dust, thereby obscuring the basic form of the underlying galaxy
4. Hubble arranged his classification of galaxies into a graphical representation known as the **tuning fork diagram** (see *Tuning Fork Diagram,* page 152)
   a. At first, astronomers believed this represented an evolutionary sequence, with galaxies beginning as ellipticals and then evolving into spirals or barred spirals
   b. Because all types of galaxies include very old stars and must be of similar age, the Hubble sequence cannot be evolutionary

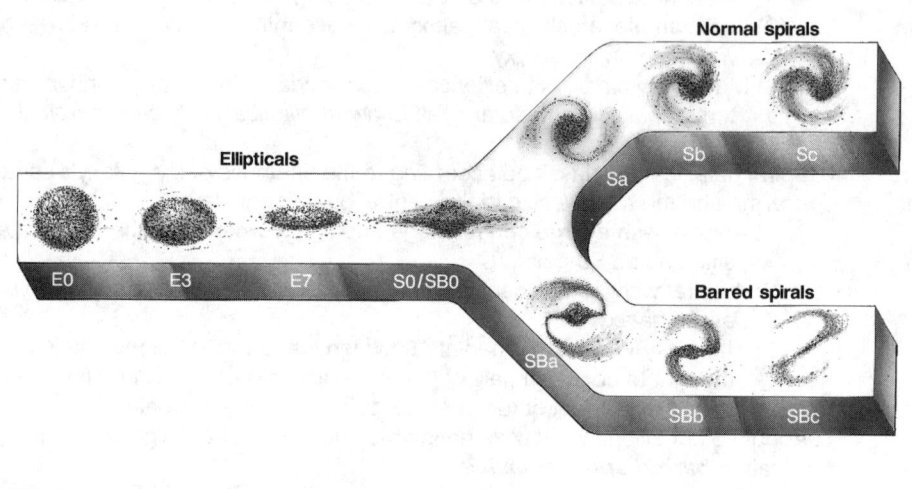

Tuning Fork Diagram

Edwin Hubble devised a way of displaying the types of galaxies—the tuning fork diagram. For quite some time, scientists believed that this was an evolutionary sequence, although the imagined direction of evolution was reversed at least once. Today, we know that in the normal course of events, galaxies do not evolve into other types of galaxies.

C. Basic properties of galaxies
1. In order to determine the fundamental properties of a galaxy, scientists must first determine the distance to it
   a. Generally, the distances to galaxies are found by identifying objects within them whose luminosity is known; such objects are called **standard candles** (see *Standard Candles and the Distance Modulus*)
   b. Cepheid variable stars are used as standard candles; they can be observed up to distances of a few million parsecs (or a few megaparsecs [Mpc])
   c. The most massive supergiant stars are found to have a common luminosity and can therefore be used as standard candles; this extends the distance scale to more than 10 Mpc
   d. Certain types of supernova outbursts have very high luminosities at peak brightness and can be used as standard candles; these can be observed as far as hundreds of Mpc away
   e. Galaxies themselves can be standard candles, if their intrinsic luminosities can be inferred
      (1) One can assume that the most luminous galaxy (or the second or third most luminous, to avoid choosing a peculiar object) in a rich cluster of galaxies always has the same luminosity
      (2) For spiral galaxies, a relationship exists between luminosity and the mass of the galaxy, which is indicated by the galaxy's rotational speed (measured from the Doppler effect in the 21-cm emission line resulting from hydrogen); this is called the *Tully-Fisher method*
      (3) For elliptical galaxies there is a similar relationship, except that these galaxies do not have a simple overall rotation; instead, scientists

Standard Candles and the Distance Modulus

As noted in Chapter 12, Measuring the Stars, the difference between the apparent magnitude (m) and the absolute magnitude (M), known as the distance modulus, gives the distance to an object. The essence of the standard candle technique is to find objects for which M is known and m can be measured from observations.

To convert the distance modulus to actual distance requires taking into account the definition of the magnitude system (a difference of one magnitude corresponds to a brightness ratio that is equal to the fifth root of 100, or a factor of 2.512) and the inverse square law. Combining these two effects yields an equation relating distance to the distance modulus, where d is the distance in parsecs and the 10 represents the standard distance used when comparing magnitudes (10 parsecs):

$m - M = 5 \log(d/10)$.

Solving this for distance yields $d = 10^{0.2(m-M)}$.

For example, suppose a supernova is seen in a distant galaxy and found to have an apparent magnitude m of +17.83 at peak brightness. The supernova is identified as Type Ia, which is known to have a peak absolute magnitude of −18.36. Inserting these values into the expression for distance, we find that d equals $10^{0.2(17.83-[-18.36])} = 10^{7.24} = 1.73 \times 10^7$ pc or 17.3 Mpc.

---

      measure the average speed of stellar motion within the galaxy to indicate the mass, hence the luminosity, and the distance
   f. The ultimate method for measuring galactic distances involves their redshifts; it is described later in this chapter
2. The masses of galaxies have a wide range of values, from about 1×10^6 to more than 1×10^{12} solar masses
   a. For spiral galaxies, the masses can be determined from the rotational speed at a large distance from the center
      (1) This is an application of Kepler's third law, analogous to its use in determining the mass of the Milky Way (see Chapter 15, The Milky Way)
      (2) As in the case of the Milky Way, this method neglects any mass farther out in the galaxy than the point at which the rotation is measured
   b. For elliptical galaxies, the mass is estimated from a measure of the average orbital speed of the stars at a known distance from the center; this is known as the *velocity dispersion,* and is measured using the Doppler effect
3. The stellar content of a galaxy can be characterized by its color and its mass-luminosity ratio
   a. The higher the content of young, hot stars in a galaxy, the more blue in color a galaxy will appear; thus, the color index is a measure of the relative numbers of Population I and Population II stars in the galaxy
   b. Because hot, young stars have high luminosities relative to their masses, the ratio of mass to luminosity is an indicator of the relative number of Population I stars in a galaxy
      (1) The mass-luminosity ratio, denoted M/L, normally is expressed in solar units; thus, a galaxy consisting entirely of Sun-like stars would have an M/L of 1, meaning that the stars were producing energy in the same proportion to their mass as the Sun does
      (2) Values of M/L range from 10 to 20 for galaxies having a high percentage of Population I stars; this means that 10 to 20 solar masses are required to produce the energy of one Sun

(3) M/L values over 100 are found in galaxies having a low percentage of Population I stars; this means that 100 to 200 solar masses are required to produce the energy of one Sun

(4) Thus, even in galaxies having a high content of hot, luminous stars, most of the mass is dim or dark

4. Systematic differences between elliptical and spiral galaxies exist
   a. Elliptical galaxies have a much wider range in luminosity and mass than spiral galaxies
   b. Elliptical galaxies have a lower percentage of Population I stars and material
      (1) Elliptical galaxies tend to be less blue and have higher mass-luminosity ratios than spiral galaxies
      (2) Elliptical galaxies generally have little or no interstellar gas and dust
      (3) Thus, elliptical galaxies tend to contain only old stars, while spiral galaxies have a combination of old and young stars

5. Elliptical and spiral galaxies are thought to be the result of differing initial conditions at the time of formation
   a. Ellipticals may have formed from relatively dense clouds of gas, which quickly formed into stars and used up all of the available gas before collapse could occur
   b. Spirals may have formed from less dense clouds, which did not complete star formation before collapsing into a disk; thus, a disk containing interstellar matter resulted, and star formation continues

D. Groups and clusters of galaxies

1. Most galaxies in the universe are members of gravitationally bound groups or clusters
2. These conglomerations range from small groups containing only a few members to rich clusters containing thousands of members
3. The Milky Way belongs to a small cluster called the **Local Group**
   a. The Local Group has about 30 known members
   b. The Milky Way and the Andromeda galaxies, two large spirals, are the dominant members
   c. The two Magellanic Clouds, satellites of the Milky Way, are our closest neighbors
      (1) These two objects, readily visible to the naked eye, lie far to the south and can be seen only from below the equator
      (2) Both are Irregular Type galaxies; the Large Magellanic Cloud is an Irregular Type I, having a hint of spiral structure
   d. Most of the galaxies in the Local Group are small, dim galaxies classified as dwarf elliptical or dwarf spheroidal galaxies
      (1) These galaxies are not much more massive than globular clusters but are far less dense
      (2) Because of their diffuse appearance and relatively low luminosities, dwarf galaxies are difficult to observe at great distances; they may be common throughout the universe, but this cannot be confirmed
4. Rich clusters of galaxies have many hundreds of members and a more uniform overall structure than the smaller groups and clusters
   a. They generally are smoothly shaped groupings, with a steady increase in the number of stars and other objects toward the center of the galaxy

b. Most rich clusters have a greater percentage of elliptical galaxies in their central regions than in the outer regions
      (1) It is thought that more spiral galaxies are formed initially
      (2) In the central, dense portion of a rich cluster, close encounters or collisions between galaxies are common
      (3) These encounters can alter a spiral galaxy, stripping away its interstellar gas and dust and converting it into an elliptical galaxy
   c. Typically, a rich cluster has a single, dominant giant elliptical galaxy at its center
      (1) These are the largest and most massive galaxies known
      (2) These galaxies often show signs of unusual, energetic activity in their cores
      (3) These central, giant elliptical galaxies may have grown as the result of mergers of galaxies, as galaxies in the central portion of the cluster have spiralled in and fallen together; this is sometimes referred to as "galactic cannibalism"
   d. Rich clusters commonly are found to have *intracluster gas* in the space between galaxies
      (1) This gas typically is very hot (approximately 1×10^8 K), and can be observed because of its resultant X-ray emission
      (2) This gas contains such heavy elements as iron, which shows that it has been affected by star formation and evolution; the gas has been pulled out of the galaxies in the cluster
   e. Masses of rich clusters are estimated from the motions of the individual galaxies in the clusters
      (1) This is analogous to the use of velocity dispersion to measure the masses of elliptical galaxies
      (2) The masses determined from this technique always are larger than the total masses of the individual galaxies plus the observed intracluster gas; this implies that rich clusters contain large quantities of dark matter
5. Clusters of galaxies are grouped into larger structures called **superclusters**
   a. Superclusters are enormous, having dimensions as large as hundreds of Mpc; they are the largest structures in the universe
   b. Superclusters are not symmetric in shape, but instead consist of sheetlike structures with large empty regions in between, giving the universe an overall cellular structure
   c. The origin of this structure is not understood, except that it is known to have developed very early in the lifetime of the universe
      (1) The origin depends strongly on the nature of the dark matter that dominates the mass content of the universe
      (2) If the dark matter is hot, meaning that its constituent particles move at speeds near that of light, gravitational clumping is unlikely
      (3) If the dark matter is cold, then early clumping is more likely; thus, the cold dark matter model appears to be favored

II. Expansion of the Universe

A. General information
1. In the 1910s and 1920s, astronomers studied the spectra of the nebulae, even though they had not yet been identified as galaxies
   a. Vesto Slipher of the Lowell Observatory noticed that the spectra of nebulae tended to be redshifted
   b. Edwin Hubble found a correlation between galaxy distance and redshift
2. In 1929, Hubble announced that the universe is expanding
   a. Each galaxy is receding from each other with a velocity (determined from the Doppler effect) that is proportional to its distance
   b. This is a uniform expansion; it would look the same to an observer anywhere in the universe because all galaxies are receding from all others as the universe itself grows in scale
3. The expansion is uniform for large-scale motions, but not for local motions, such as galactic orbits within a cluster of galaxies
   a. The center of mass of a cluster of galaxies moves with the general expansion of the universe, while individual galaxies follow their own orbits about the center of mass
   b. As a result of such local motions, the Andromeda galaxy actually is approaching the Milky Way instead of receding from it
4. The expansion of the universe had been anticipated by some astronomers who had applied the laws of general relativity to the universe and found that it could not be static (see Chapter 17, Cosmology)

B. The Hubble constant
1. Hubble found that the recession velocity of a galaxy is proportional to its distance
   a. The expansion can be expressed by an equation known as the **Hubble law;** it is $v = Hd$, where v is the recession velocity of a galaxy, H is a number called the **Hubble constant,** and d is its distance; if v is in units of km/sec and d in Mpc, then H has units of km/sec/Mpc
   b. The value of H has extremely important implications for the overall nature of the universe, but also is difficult to measure accurately
2. On a plot of recession velocity versus distance, the points for individual galaxies fall along a diagonal line (see *A Modern Version of the Hubble Law*)
   a. The slope of this line, which gives the rate at which velocity increases with distance, is equal to the Hubble constant
   b. The measured value of H has varied considerably, and even today astronomers do not agree on the value
      (1) Hubble initially found that the value of H was roughly 540 km/sec/Mpc; that is, for every Mpc of distance, the recession velocity increases by 540 km/sec
      (2) Hubble's value was far too high, as a result of errors in measuring galactic distances; modern values range between 45 and 90 km/sec/Mpc

C. The age of the universe
1. The age of the universe can be estimated from the value of the Hubble constant
2. This assumes that the universe has expanded to its present size from a single point; in effect, we calculate how long it has taken for the galaxies to reach their present positions at their present rate of expansion

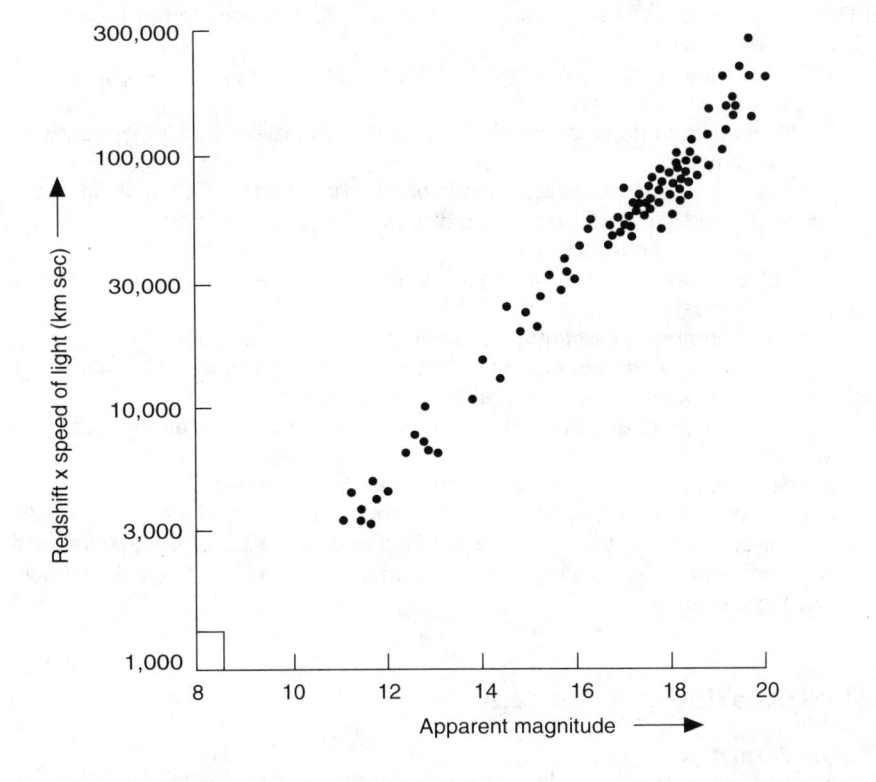

A Modern Version of the Hubble Law

Apparent magnitude may appear on the horizontal axis instead of distance because the two quantities are related, especially if the diagram is limited to galaxies of the same type, as this one is. The small rectangle at lower left indicates the extent of the relationship as Hubble first knew it; today many more galaxies, much dimmer, are included. The most rapid galaxies are not actually traveling at the speed of light as this diagram implies; relativistic corrections have not been applied.

3. Because time equals distance divided by velocity, the age of the universe equals distance (d) divided by velocity (Hd); age then equals d/Hd or 1/H
   a. To calculate the age of the universe, it is necessary to convert Mpc to km; then the km units cancel out and the result will be in seconds
   b. If H = 540 km/sec/Mpc as Hubble originally thought, the age comes out to just under 2 billion years, which is less than the known age of Earth
   c. A modern value of H = 45 km/sec/Mpc yields an age of 22 billion years, comfortably longer than the estimated ages of the globular clusters, the oldest known objects
4. Ages calculated in this manner are overestimated because the expansion rate has not been constant; it was more rapid in early times than it is today

D. The use of redshifts to measure distances
1. If it is assumed that the observed redshift of a galaxy is the result of the expansion of the universe, the distance can be calculated using the Hubble law
   a. Solving the Hubble law for distance yields $d = v/H$, where d is the distance (in Mpc), v is the velocity (in km/sec), and H is the Hubble constant (in km/sec/Mpc)
   b. Thus, all that is needed is to measure v from the Doppler effect and to know the value of H
2. This technique allows the measurement of galactic distances to the edge of the observable universe
   a. Distances as large as thousands of Mpc, or billions of parsecs, can be measured
   b. This technique depends on all those described previously for measuring distances
      (1) The use of redshifts depends on knowing the Hubble law, which depends, in turn, on galactic distances measured through the use of various standard candles
      (2) Standard candles, in turn, depend on distance measurements to individual stars
      (3) Ultimately, all distance-determination techniques depend on the measured value of the distance between the Sun and Earth because this is the basis for stellar parallax measurements
3. Observing objects at great distances means that we see them as they were long ago
   a. The "look-back" time is equal to the distance in light-years
   b. Thus, galaxies seen at distances of billions of parsecs, which corresponds to billions of light-years (recall that one parsec equals 3.26 light-years), are seen as they were billions of years ago; in a sense, we are looking backward in time

III. Active Galaxies and Quasars

A. General information
1. In the early days of radio astronomy (in the 1930s and 1940s), astronomers discovered that the brightest radio sources in the sky are galaxies
   a. These **radio galaxies** emit far more energy at radio wavelengths than at visible wavelengths
   b. The presence of such strong radio emission indicates that some unusual energetic activity must be taking place
2. Subsequently, other forms of galaxies with energetic activity also were found; collectively, these are called **active galactic nuclei (AGNs)**
3. In 1960, a type of starlike object was found that emits strongly at radio wavelengths; these became known as **quasi-stellar objects (QSOs)** or, more commonly, **quasars**
4. Today, it is understood that all of these objects have features in common, and may have similar origins

B. Radio galaxies
1. Radio galaxies typically are ellipticals, frequently the giant ellipticals found at the centers of rich clusters of galaxies
   a. These are galaxies thought to have formed from the merger of several smaller galaxies
   b. Generally, these elliptical galaxies have peculiar appearances, such as the jet of ionized gas protruding from the galaxy called Virgo A (also known as M87) or the belt of interstellar gas and dust across the galaxy known as Centaurus A
2. The radio emission invariably comes not from the visible galaxy, but from pairs of radio-emitting lobes on opposite sides of the visible galaxy
   a. These radio lobes are enormous and normally are much larger than the visible galaxy
   b. Evidently these lobes consist of ionized gas that has been ejected from the central galaxy along opposing jets
3. The radio emission is produced by the synchrotron process (see Chapter 14, Stellar Remnants), which requires a strong magnetic field and high-energy particles (typically electrons) moving at nearly the speed of light

C. Seyfert galaxies and BL Lacertae objects
1. In the 1930s, American astronomer Carl Seyfert found several spiral galaxies with unusually bright, blue nuclei
2. Now known as Seyfert galaxies, these were found to have emission-line spectra
   a. In some cases, the emission-line widths imply, through the Doppler effect, local gas motions of several thousand km/sec
   b. The degree of ionization represented by the emission lines indicates local gas temperatures as high as 1 million K
3. Seyfert galaxies frequently vary in brightness with short timescales
   a. Sometimes variations are seen every few days
   b. These rapid variations imply that the size of the emitting region must be very small; in order to change in brightness, a volume of gas cannot be larger than the distance light can travel in the time it takes for the variation to occur, which implies that the emitting region in Seyfert galaxies can be no larger than a few light-days across
   c. Thus, Seyfert galaxy nuclei are regions where immense quantities of energy are emitted from a small volume, suggesting that black holes may be responsible for this effect
4. About 10% of Seyfert galaxies also are strong radio sources
5. Another class of galaxies with bright nuclei are the **BL Lacertae objects,** named after the first one found, which was originally thought to be a variable star
   a. The nucleus is so bright relative to the surrounding galaxy that it took years before observations were sensitive enough to reveal that this type of galaxy existed
   b. BL Lacertae objects are active galaxies whose nuclei beam radiation directly toward Earth

D. Quasars
1. The first quasars were discovered because they were radio emitters having a starlike appearance

2. Observations at visible wavelengths revealed these to be emission lines, with spectral lines that were at first unidentifiable and later found to be enormously redshifted lines of ordinary gases
   a. In the first two quasars discovered, 3C48 and 3C273, the recession speeds were 30% and 15% of the speed of light, respectively
   b. The redshifts commonly are expressed in terms of the shift in wavelength relative to the rest position of the spectral line; in equation form, the redshift, which is analogous to the Doppler shift (z), is $z = \Delta\lambda/\lambda$, where $\Delta\lambda$ is the shift in wavelength and λ is the rest wavelength of the line
   c. Redshifts as large as 4.73 have been found; for example, the strong ultraviolet line of hydrogen, whose rest wavelength is 122 nm, is shifted to 699 nm, in the red part of the visible spectrum
   d. These large redshifts do not mean that the quasars are moving away from us faster than the speed of light, as the normal Doppler formula would imply (see Chapter 4, Light and the Atom); instead, the relativistic Doppler formula must be used
      (1) The relativistic Doppler formula, where v is the velocity of recession, c is the speed of light, and z is the redshift, is:
      $$v = c\left[\frac{(z+1)^2 - 1}{(z+1)^2 + 1}\right]$$
      (2) For example, if z = 4.0, the speed is found to be 0.923 c; that is, a quasar having a z of 4.0 is travelling at 92.3% of the speed of light, not 4 times the speed of light as the normal Doppler shift formula would imply
3. Once the enormous redshifts of quasars were recognized, it became apparent that the quasars must be extremely distant, more distant than any known galaxies
   a. This implies, in turn, that quasars are far more luminous than any other known objects; they have up to 1,000 times the luminosity of bright galaxies
   b. This also implies that the quasars existed only at early times in the history of the universe, because they invariably have large redshifts and therefore are seen as they were long ago
4. Many quasars have emission and absorption lines in their spectra, with different redshifts
   a. The emission redshift always is greater than the absorption redshifts, indicating that the emitting region is farther away than the absorbing regions
   b. Absorption lines may form in gas clouds that lie between us and the distant quasars
      (1) These absorbing clouds may represent clouds in intergalactic space that never formed into galaxies
      (2) In some cases, the absorbing clouds appear to be the extended halos of galaxies lying along the lines of sight to quasars
5. Many quasars are found to vary in brightness, with changes occurring every few days
   a. This implies that the emitting region is very small, similar to the emitting regions in Seyfert galaxies
   b. Once again, enormous quantities of energy come from very small volumes, suggesting that black holes are the source of this energy

6. Scientists believe quasars are the nuclei of young galaxies, seen as they were billions of years ago
   a. This suggestion arises in part from the many similarities between quasars and AGNs
   b. Surrounding galaxies are observed around all quasars nearby enough that a galaxy could be seen; that is, in every case where the parent galaxy could be detected, it has been
7. The cores of these young galaxies may be supermassive black holes
   a. These black holes have masses as large as 100 million solar masses
   b. The immense energy production from these galactic nuclei may arise from accretion disks around the black holes
   c. The jets of superheated gas from the nuclei may emerge along the rotational axes of the accretion disks
      (1) The jets are produced by a combination of rotational and magnetic forces
      (2) These jets are responsible for the opposing radio-emitting lobes in radio galaxies as well as the beams of radiation from BL Lacertae objects
   d. Modern galaxies that have active nuclei may be the descendants of quasars; their central activity has diminished as the amount of matter falling into the accretion disk has diminished
8. Evidence suggests that quasars formed early in the universe from the mergers of young galaxies
   a. Collisions between galaxies were more common then because of the higher density of the universe at earlier times in its expansion
   b. Observations have revealed a higher frequency of collisions and mergers among very distant galaxies at early times in the universe

Study Activities

1. Explain why the Hubble sequence of galaxy types (the tuning fork diagram) is not a sequence of galaxies that evolve from one type into the next.
2. Explain how the masses of galaxies are measured; summarize the similarities with the measurement of stellar masses (described in Chapter 12).
3. Calculate the age of the universe if the value of the Hubble constant is 100 km/sec/Mpc. Would this be consistent with the ages of globular clusters?
4. Find the distances to the following galaxies and quasars, assuming that the value of the Hubble constant is 50 km/sec/Mpc:
   (a) a galaxy having a redshift of 0.0047
   (b) a galaxy having a redshift of 0.457
   (c) a quasar having a redshift of 2.674.
   (Note that first you must convert each redshift in terms of recession velocity; the relativistic Doppler formula can be used in all cases, and must be used for z greater than 0.1.)
5. Explain how a single quasar can have simultaneous emission lines and absorption lines at different redshifts.

17

Cosmology

Objectives

After studying this chapter, the reader should be able to:
- Explain the methods astronomers use to probe the overall structure and evolution of the universe.
- Describe the basic predictions of general relativity for the future of the universe.
- Explain how observations are used to test the predictions of general relativity.
- Summarize the properties of cosmic background radiation.
- Describe the processes that occurred during the earliest phases of the universe's expansion.
- Explain the inflationary universe model and its relationship to the big bang model.
- Summarize the evolution of the universe from the earliest moments to the present.

I. The General Relativistic Framework for Cosmology

A. General information
1. The discovery of the expansion of the universe gave the first real evidence that the universe changes with time and may have had a definite beginning
2. *Cosmology* is the term generally applied to any study of the universe as a whole
   a. Technically, cosmology refers only to the description of the universe as it is today
   b. *Cosmogony* is the term for study of the formation and evolution of the universe; the term cosmology commonly is used to mean cosmogony
3. Until the early 20th century, most cosmology consisted only of philosophical speculation, with little basis in mathematical models

B. Albert Einstein's theory of general relativity
1. In the 1910s, Albert Einstein developed a mathematical description of nature in which gravity is viewed as a geometric effect
   a. He postulated that light always takes the shortest path between two points
   b. Because the path of light passing near a massive object follows a curved path caused by gravitational attraction of the nearby mass (as seen by a distant observer), space itself is curved in the presence of mass
2. Using this formalism to describe gravity, Einstein developed a set of equations, called the *field equations,* describing the interaction of matter, energy, and gravity in the universe
3. The task of developing a mathematical model of the universe then became one of finding solutions to the field equations

a. One common assumption that often is made is that the universe is *homogeneous,* meaning that it looks the same to observers at all locations and there is no special location within the universe that is preferred over other locations in any way
b. Another common assumption is that the universe is *isotropic,* meaning that to an observer at any location it looks the same in all directions
   (1) The statement that the universe is homogeneous and isotropic sometimes is called the **cosmological principle**
   (2) Large-scale modern observations appear to confirm this principle
c. Einstein's initial solution to the field equations showed that the universe could not be static, but instead that it must expand or contract
   (1) Einstein was inclined to believe that the universe must be static, so he introduced an arbitrary force called the **cosmological constant** in order to keep the universe from collapsing
   (2) When Edwin Hubble discovered the expansion of the universe, Einstein referred to his introduction of the cosmological constant as the worst blunder he ever made
   (3) Today, most cosmologists assume that there is no cosmological constant; because observations do not rule out the possibility, a term for it—which equals zero—generally is included in the equations
d. In the early 1920s, Belgian cleric and scientist Georges LeMaître developed a solution to the field equations in which the universe was expanding from a state of high density and temperature; this became known as the **big bang theory** of the universe
4. The big bang model allows for three possible outcomes of the expansion of the universe
   a. The expansion may continue forever, slowing but never coming to a halt
      (1) This is called the *open universe* model
      (2) In this model, the universe has no boundaries, is infinite in size, and is said to have negative curvature; an analogy would be a saddle-shaped surface
   b. The expansion may eventually stop and reverse, starting a universal contraction
      (1) This is called the *closed universe* model
      (2) In this model, the universe has no boundaries, but is finite in size and is said to have positive curvature; an analogy is the surface of a sphere
   c. The expansion may slow at such a rate that it approaches a stop, but it requires an infinite amount of time to do so
      (1) This is called the *flat universe* model
      (2) In this model, the universe has no boundaries, is infinite, and has no overall curvature; an analogy would be a flat plane
5. Modern cosmology is devoted to determining which of these three models is correct

II. Observational Cosmology

A. General information

1. There are two categories of observational tests that can be applied to the big bang theory: those that the test the theory in general and those attempting to determine which of the three forms describes reality
2. To test whether the big bang model is correct, one must look for consequences that would not have developed under alternative models—for example, conditions or artifacts left over from the early times when the universe was hot and dense
3. To discriminate among the three possible outcomes, the normal technique is to try to determine the values of certain mathematical constants from field equations which should be significantly different for the different scenarios
   a. One such numerical factor is the Hubble constant (H), which describes the rate of expansion of the universe (see Chapter 16, Galaxies); this helps to determine whether the expansion will stop or continue forever
   b. Another important factor is q, the *deceleration parameter*, which measures how much the expansion has slowed and suggests whether it will eventually stop
   c. Another key parameter is T, the age of the universe, which is related to the question of outcome by virtue of setting a timescale
   d. Another factor is a measure of the average density of the universe (designated Ω); the density governs whether or not there is enough mass to counteract the momentum of expansion
   e. A final factor is the cosmological constant (Λ) mentioned earlier, which is the term introduced by Einstein when he erroneously thought the universe must be static; even though the universe is dynamic, the possibility exists that some previously unknown force affects the expansion and its outcome

B. Cosmic background radiation

1. George Gamow realized some 50 years ago that if the big bang model was correct, the early universe must have been very hot and dense
   a. Gamow was interested in this primarily because he was exploring the possibility that nuclear reactions initiated during the big bang might have produced all the observed heavy elements
   b. A hot, dense early universe should glow as a result of thermal radiation, and that radiation should still fill the universe today
      (1) The radiation would have been redshifted by the expansion of the universe, thereby reducing its temperature
      (2) Gamow expected that the temperature of the radiation should be some 25 K today; thus, the wavelength of maximum emission should have been around 1 cm (see the description of Wien's law in Chapter 4, Light and the Atom)
      (3) At the time, no one attempted to search for this cosmic radiation
2. In the 1960s, Robert Dicke independently predicted that the universe should be filled with remnant radiation from the early stages of the big bang expansion
   a. Dicke and collaborators began to build an instrument to detect the remnant radiation
   b. Such a detection is made difficult by the fact that Earth's atmosphere glows in the same wavelength region as the peak of the expected remnant radia-

tion, thus obscuring all remnant radiation except for the short-wavelength side of the expected peak
3. In 1965, radio astronomers Arno Penzias and Robert Wilson discovered background noise in a new radio telescope; this noise turned out to be **cosmic background radiation**
4. The radiation was found to match a thermal spectrum for a body surface temperature of 2.7 K; hence, the radiation is sometimes called "three-degree background radiation" (Gamow's predicted temperature was higher because of uncertainties in his big-bang model)
5. In order to demonstrate that the radiation truly originated from the big bang, scientists had to demonstrate that its spectrum is precisely thermal and that the radiation is isotropic (the same in all directions)
   a. Early measurements indicated that the spectrum of the radiation fits a thermal spectrum, but measurements on the short-wavelength side of the peak (which turned out to be close to 1 mm in wavelength) were made difficult or impossible by Earth's atmosphere
   b. Space-based observations, particularly those made by the *Cosmic Background Explorer (COBE)* satellite in the early 1990s, have measured the short-wavelength side of the peak accurately
   c. The *COBE* results have yielded a precise temperature of 2.735 K for the radiation.
   d. Radiation is isotropic on the largest scales
      (1) However, there is a small Doppler shift of the radiation as measured in opposite directions as a result of Earth's motion relative to the rest wavelength of the radiation; this *dipole anisotropy* does not reflect a true departure of the radiation's uniformity with respect to direction
      (2) Scientists use dipole anisotropy to deduce the motion of Earth relative to the rest wavelength of the radiation; this motion actually is a combination of Earth's spin and orbital motions within the solar system, the solar system's orbital motion in the galaxy, the galaxy's motion within the **Local Group,** and the motion of the Local Group within the local supercluster of galaxies to which it belongs
      (3) The *COBE* observations revealed subtle regional variations in the temperature of the cosmic background radiation, indicating that the universe had developed some structure at a very early time; this helps to pinpoint the era when the formation of superclusters of galaxies began
6. Thus, the properties of the cosmic background radiation strongly support the big bang model of the universe, having matched all of the predictions of that model

C. Observational clues to the fate of the expansion
1. Of the five numerical parameters that describe the expansion of the universe, two are particularly useful in determining whether the universe is open, closed, or flat
2. One of these two is the density parameter Ω representing the ratio of the actual density to the *critical density,* which is the density required to reverse the expansion
   a. The value of the critical density depends on the value of H, the Hubble constant, and it is about 1×10^{-29} g/cm^3 (or the equivalent of one hydrogen atom for every 1×10^5 cm^3)

b. The observed density is determined by attempting to add up the mass of all the galaxies in some well-defined, large volume of space; the result falls far short of the critical density
   c. The unknown quantity of dark matter in the universe is crucial in determining the value of Ω; it appears that the dark matter may bring the actual density close to the critical density, suggesting that the universe is either flat or nearly flat
3. The other useful factor in determining whether the universe is open, closed, or flat is the deceleration parameter (q)
   a. One way to attempt to measure q is to compare the expansion rate at early times with the present-day value of H
      (1) In principle, the expansion at early times can be determined by extending the Hubble expansion law to the faintest and most distant galaxies; because of the "look-back in time" effect, this would yield the expansion rate early in the history of the universe
      (2) This has yielded ambiguous results because it is very difficult to accurately measure the distances to the most faraway galaxies and establish their expansion rates
   b. Another way to measure q indirectly is to measure the abundances of elements thought to have formed during the early expansion because these abundances depend on how long it took for the universe to expand and cool beyond the point where nuclear reactions could continue
      (1) In order to do this, it is most useful to consider elements that are not formed in stellar nuclear reactions, so that the quantity we observe today is the quantity produced in the big bang; one such species is deuterium, a form of hydrogen having one proton and one neutron
      (2) Again, ambiguous results are found, largely because of the possibility that some quantities of deuterium may be produced by stars; taken at face value, the observed abundance of deuterium indicates that the universe is open
   c. On balance, deceleration measures seem to indicate an open universe, but this is not considered conclusive because of the uncertainties

III. Evolution of the Universe

A. General information
1. Astronomers can derive surprisingly accurate and complete information about the big bang
2. Theoretical studies of the early universe are relatively straightforward because the universe had a simpler structure and its curvature was less important in the early stages
3. Experimental data from particle accelerators produces conditions similar to the early universe
4. The key processes that occur under high-energy conditions are particle pair formation and particle annihilation

B. Particle Formation and Annihilation
1. Both processes are the result of the relationship between mass and energy, expressed by Einstein's equation $E = mc^2$

2. Pair formation occurs when a photon or a pair of photons of radiation spontaneously form a pair of particles
   a. The total mass of the particles equals the energy of the photon or photons, through the equation $E = mc^2$
   b. The particles produced always represent a particle-antiparticle pair having equal masses but being opposite in other properties, such as electrical charge
   c. A pair of photons is required for particle pair formation in a vacuum; in this case, the sum of the photon energies equals the rest mass energy of the pair of particles that become a pair
   d. A single photon can form a particle pair if other nuclei are present to absorb some of the momentum of the photon; in this case, the photon's energy must equal the total rest mass energy of the particle pair that is created
3. Particle annihilation occurs when a particle-antiparticle pair merge, converting their total mass into a photon whose energy is given by the equation $E = mc^2$
4. As the early universe expanded and cooled, the energies of the photons and particles filling it diminished; thus, these high-energy processes occurred only for a short time

C. The early expansion
1. Modern cosmological theory cannot state precisely how and why the universe began from a vacuum (although progress toward this goal is being made); it is assumed that random energy fluctuations in a localized region attained very high density and temperature, and that the universe expanded from that localized region
2. At the earliest moment, the four natural forces were unified into one
   a. The four forces, in order of weakest to strongest, are gravity, the electromagnetic force, the weak nuclear force, and the strong nuclear force
   b. The four forces today are distinguished from one another by their differing interactions with different types of particles
      (1) *Gravity* is the attractive force between any two masses
      (2) The *electromagnetic force,* which can be an attractive or a repulsive force, acts on electrically charged particles and is inversely proportional to the square of the distance between them, as is gravity
      (3) The *weak nuclear force* comes into play only in certain types of nuclear reactions
      (4) The *strong nuclear force* acts only over very small distances and is responsible for holding the particles of an atomic nucleus together
   c. When the universe was 1×10^{-43} seconds old, the temperature was 1×10^{32} K and the four forces were indistinguishable from one another; at this time, it is said that the four forces were unified; astronomers use time (t) to refer to the age of the universe
   d. First gravity (at $t = 1 \times 10^{-43}$ seconds), then the strong nuclear force (at $t = 1 \times 10^{-35}$ seconds), and finally the electromagnetic and weak nuclear forces (at $t = 1 \times 10^{-12}$ seconds) became separate
3. At $t = 1 \times 10^{-12}$ seconds, the temperature was 1×10^{15} K and the major particles that were appearing and annihilating were *quarks,* the most fundamental of all elementary particles

4. At t = 1×10^{-6} seconds, the principal particles in the universe were protons, neutrons, electrons, and neutrinos
   a. The protons and electrons, along with their antiparticles, were in equilibrium with the radiation field; that is, they were formed and destroyed at equal rates
   b. Protons and antiprotons formed spontaneously from photons, and then annihilated each other, producing more photons
   c. The same happened for electrons and *positrons* (also known as anti-electrons)
   d. Neutrons do not form pairs readily because they are electrically neutral and do not interact strongly with radiation; most neutrons formed from a reaction between protons and electrons, called *inverse beta decay,* which also produces neutrinos
      (1) Neutrinos interact with other matter only through the weak nuclear force, which means that they can pass through normal matter without interaction most of the time
      (2) Thus, the universe has had a high density of neutrinos throughout its evolution
5. As the universe cooled further, the temperature dropped below the point where the photons had enough energy to form particle-antiparticle pairs
   a. The remaining particles and antiparticles continued to annihilate each other
   b. There would be no matter in the universe today, except for a slight imbalance between the numbers of particles and antiparticles that existed then; this imbalance amounted to about one extra particle for every billion particle-antiparticle pairs
      (1) The reason for the asymmetry and the dominance of matter over antimatter is not well understood
      (2) One possibility is that the decay of some of the earlier particles, those that formed and ceased to exist before the era of protons and electrons, may have been asymmetric in the sense that it resulted in slightly more particles of ordinary matter than antimatter
6. During the next 3 or 4 minutes, nuclear fusion reactions created lightweight elements
   a. These reactions were possible only because some neutrons survived from earlier times
      (1) Free neutrons are not stable, and normally decay into protons and electrons within 15 minutes
      (2) Free neutrons were necessary for the first fusion reactions
   b. The first nucleus formed (after the proton) was that of *deuterium,* the heavy form of hydrogen consisting of one proton and one neutron
   c. Deuterium nuclei were then able to capture additional neutrons, forming *tritium,* another form of hydrogen containing one proton and two neutrons
   d. Next, tritium captured an additional neutron, forming helium (two protons and two neutrons; one of the captured neutrons emits an electron and turns into a proton)
   e. Deuterium also reacted with a proton, to make ^{3}He (helium) or with another deuterium nucleus to make ^{4}He; other reactions also are possible
   f. The reactions produced roughly 20% (by mass) of helium nuclei during this time; nearly all the rest was still in the form of hydrogen nuclei (protons), with traces of other lightweight elements, such as lithium and beryllium

7. After the nuclear reactions stopped, nothing of great significance happened for nearly a million years, until the time when electrons combined with nuclei to form atoms
   a. This process, called *recombination,* occurred when the temperature of the universe was about 3,000 K
   b. By this time, the radiation had cooled (that is, it had been redshifted by the universe's expansion) so that its wavelength of peak emission was in the visible to ultraviolet portion of the spectrum
8. Once recombination into neutral atoms had occurred, the universe became transparent, as matter and radiation stopped interacting strongly with each other
   a. This event is referred to as the *decoupling of matter and radiation*
   b. Since that time, the radiation has continued to cool (and continues to be redshifted by the universe's expansion), and is observable today as cosmic microwave background radiation having a temperature of 2.735 K
   c. Since the decoupling, the matter in the universe has become organized into stars and galaxies
      (1) At or before the time of decoupling, the universe already had developed density variations, evidenced by intensity variations in the the microwave background radiation
      (2) These early density fluctuations probably also were reflected in the distribution of matter at the time of decoupling, and may have formed the precursors to superclusters of galaxies, whose origin is otherwise difficult to explain
9. All elements other than hydrogen, helium, and traces of lithium and beryllium have been formed in stellar interiors (see Chapter 13, Stellar Structure and Evolution), and substantial quantities of helium have been added by stellar nuclear reactions

D. The inflationary universe
1. A recently developed modification of the big bang theory describes a sequence of events that might have happened at very early times
   a. These events would have occurred before the main steps described in the big-bang expansion
   b. In a sense, this new theory describes a larger picture of our universe with the big bang being a subordinate part
2. In this theory, the universe is thought to have undergone a very short period of extremely rapid expansion at very early times
   a. The rapid expansion phase has been likened to the sudden appearance of bubbles in water that has been supersaturated with a gas, such as carbon dioxide; hence, the name **inflationary universe theory**
   b. In a time of 1×10^{-30} seconds, the universe may have expanded by a factor of 1×10^{25} in size
3. This rapid expansion may have been triggered by a special state called a *false vacuum* that would have existed at very early times as a result of the extremely high-energy density of the universe
   a. If matter is present in a false vacuum, gravity acts as a repulsive force, thereby causing expansion as particles repel each other
   b. Thus, if particles are created from pair production at a time when false vacuum prevails, the appearance of the particles forces rapid expansion to occur

c. As the expansion proceeded, the universe cooled, and in time (about 1×10^{-30} seconds), the false vacuum would have been broken and the expansion would have slowed
4. Following the rapid expansion phase, the "normal" expansion would have continued as described earlier
5. The inflationary universe model is supported by observations
    a. The inflationary universe model predicts that the universe is flat, and this is consistent with observations
        (1) As noted earlier, the mass density of the universe, allowing for dark matter, is very close to the critical density
        (2) Prior to the development of the inflationary universe theory, there was no reason to expect the universe to be flat or even nearly flat
    b. The inflationary universe model also predicts that the universe should be uniform on a large scale
        (1) Before the initial rapid expansion phase, the universe was small enough that physical conditions would have been kept in equilibrium
        (2) Again, before this model was developed, there was no natural explanation for the overall uniformity of the universe
6. The inflationary universe model allows for the possibility that many other universes may exist
    a. New universes may be created whenever random fluctuations in the energy density create local regions of false vacuum
    b. Such conditions are extreme and therefore unlikely to occur as a result of random fluctuations, but given enough time and space, even very unlikely events can happen
    c. If new universes were created in this way, it is not clear that any form of communication between them and our universe would be possible, nor would we have any way of knowing that other universes exist

E. The future of the universe
1. If the universe is closed, the expansion will be reversed, and the universe will end in a new state of high density and high temperature
    a. This would take tens or hundreds of billions of years
    b. The steps described in the previous section would be reversed, in the sense that matter would be broken down into elementary particles and then radiation, and the four forces would be reunified
    c. Some cosmologists have speculated that this would be part of a cyclical universe; that a new big bang would follow, and the cycle would repeat itself
2. If the universe is open or flat, the expansion will continue indefinitely, and the matter and energy content of the universe will gradually dissipate
    a. The temperature of the universe will continue to drop, approaching absolute zero (0 K)
    b. Stellar nuclear processes will continue to convert hydrogen into heavier elements and stellar remnants; it is estimated that all the hydrogen will be used up by age 1×10^{14} years
    c. More and more of the mass of the universe will be converted into such remnants as white dwarfs, neutron stars, and black holes
    d. All planets will eventually (by age 1×10^{17} years) be lost from their parent stars as a result of random collisions between stars
    e. Similarly, galaxies will dissipate by age 1×10^{18} years

f. Protons are thought to be so unstable that they will disintegrate spontaneously in an average time of 1×10^{32} years; by this time, the universe will consist only of free positrons, electrons, black holes, and radiation

g. Even black holes can decay — quantum mechanical processes exist whereby particles can escape from them; by age 1×10^{100} years, the universe will contain nothing but positrons, electrons, and radiation

Study Activities

1. Summarize the assumptions about cosmology based on general relativity; list what conditions are assumed to prevail in the universe. Explain how these assumptions are related to observational information gained from previous chapters, particularly the description of the distribution of matter in Chapter 16.
2. Summarize the contrasts among the three fates of the universe that are derived from general relativistic cosmology (the big bang theory).
3. Explain each of the five numerical factors that characterize the evolution of the universe in the big bang model.
4. Summarize the arguments supporting the conclusion that observed cosmic background radiation is the remnant radiation resulting from the big bang.
5. Describe the observational tests that can be made to determine which of the three outcomes of the big bang expansion will occur. What are the current results of these tests?
6. Explain how particles and radiation are converted from one form to the other, and how this is related to the temperature of the universe at different times during the early expansion.

Appendix

Selected References

Index

Appendix: Glossary

Absolute magnitude—measure of the luminosity of a star, defined as the magnitude the star would have if it were located 10 parsecs from the Earth; represented by M

Absorption lines—dark spectral lines appearing at specific wavelengths in an otherwise bright spectrum

Accretion disk—gravitationally compressed disk of inward-falling matter surrounding a massive, generally compact, central object

Active galactic nucleus (AGN)—luminous, compact nucleus found in certain galaxies, typically characterized by an emission-line spectrum and time variability

Albedo—fraction of light reflected from a planet or body in the solar system

Alpha capture reaction—reaction in which an alpha particle is captured by nuclei to build up heavier nuclei in evolved stars

Apparent magnitude—observed magnitude of a star or other object; represented by m

Asteroid—small, Sun-orbiting body that may be residual debris left over from the formation of the solar system; most asteroids orbit the Sun between Mars and Jupiter

Astrometric binary—double star system identified as a binary because one or both stars are observed to undergo periodic shifts in position as a result of orbital motion

Astrometry—science of measuring stellar positions

Astronomical unit (AU)—unit of distance equal to the average distance between the Sun and Earth; 1 AU equals 1.4959787×10^8 km

Barred spiral galaxy—see *Spiral galaxy*

Big bang theory—any model of the early history of the universe positing that the universe began from a small point and has been expanding since that time

Binary stars—two or more stars that are gravitationally bound to each other and orbit one another

Blackbody radiation—see *Thermal radiation*

Black hole—object that has collapsed under its own gravitation to such a small radius that light cannot escape its gravitational field

BL Lacertae object—type of active galaxy with a bright nucleus that beams radiation directly toward the Earth

Celestial sphere—imaginary sphere formed by the sky used as a convenient device for defining the positions of objects on the sky; the *celestial poles* are the points directly over the north and south poles of the Earth, and the *celestial equator* is the projection of the Earth's equator onto the sky

Center of mass—in any system of two or more objects, the point that moves through space uniformly as the individual objects orbit that point; the balance point in an isolated system

Cepheid variable—class of variable star in which the period of variation is proportional to the luminosity; used as a distance indicator

Chromosphere—a thin layer of hot gas (typically 10,000 to 20,000 K) that lies above the photosphere of the Sun and other cool stars

CNO cycle—dominant hydrogen-fusing reaction for upper-main sequence stars

Color index—measure of stellar surface temperature, defined as the difference between the *blue (B) magnitude* and the *visual (V) magnitude;* the value of (B − V) is negative for hot stars and positive for cool ones

Color-magnitude diagram—Hertzsprung-Russell diagram for a cluster of stars, in which apparent magnitude (m) is plotted against color index (B − V)

Convection—process in which heat energy is transported by motion, such as atmospheric circulation or boiling

Corona—very hot, extended outer region of gas surrounding the Sun and other cool stars

Cosmic background radiation—primordial radiation field that fills the universe as a remnant of the very intense, high-energy radiation that characterized the universe at very early times; the temperature of the radiation today is 2.735 K

Cosmic rays—highly energetic atomic nuclei travelling through space; most come from the Sun, but some come from galactic sources, such as supernova explosions

Cosmological constant—term added to the field equations of general relativity by Albert Einstein in order to allow mathematical solutions supporting the theory that the universe is static (neither expanding nor contracting)

Cosmological principle—assumption of cosmological theory that the universe is homogeneous (has similar structure and content everywhere) and isotropic (appears the same in all directions)

Cosmology—study of the universe as a whole

Dark matter—nonluminous matter thought to comprise 80% to 90% of the universe

Declination—equatorial coordinate specifying angular distance from the celestial equator

Degenerate electron gas—see *Degenerate gas*

Degenerate gas—highly compressed gas in which subatomic particles (generally electrons or neutrons) are packed together as closely as allowed by the laws of quantum mechanics; the resistance of the particles to further compression creates a form of pressure which is independent of temperature and is sufficient to prevent the mass of a star (white dwarf or neutron star) from collapsing

Degenerate neutron gas—see *Degenerate gas*

Differential gravitational force—see *Tidal force*

Differentiation—sinking of relatively heavy elements toward the center of a star or planet; this process occurs only if the body is fluid

Distance modulus—measure of the distance to an object, defined as the difference between its apparent and absolute magnitudes (m−M)

Diurnal motion—any apparent motion in the sky that results from Earth's rotation

Doppler effect—shift in wavelength or frequency of a wave due to the relative motion of the source and the observer

Eclipsing binary—double star system in which the two stars alternately pass in front of each other, as seen from Earth, causing periodic dimmings in the total brightness of the system's ecliptic

Ecliptic—plane of Earth's orbit, or the apparent annual path of the Sun across Earth's sky

Effective temperature—surface temperature of a blackbody radiation emitter having the same luminosity and size as a known, measured star

Electromagnetic radiation—radiation of varying wavelengths; this includes visible radiation (light) and invisible radiation

Electromagnetic spectrum—range of all wavelengths of radiation (including invisible wavelengths)

Electron—negatively charged particle that orbits an atomic nucleus

Elliptical galaxy—spherical or spheroidal galaxy having no disklike component or spiral structure

Emission lines—bright spectral lines appearing at specific wavelengths in the electromagnetic spectrum

Emission nebula—see *H II region*

Equatorial coordinate system—most commonly used system for measuring locations in the sky, consisting of right ascension and declination coordinates

Equinox—either of the two points on the sky where the celestial equator intersects the plane of the Earth's orbit; at the time when the Earth passes through one of these points, the Sun appears directly over the equator and the lengths of day and night are equal

Escape speed—upward speed a body must have so that it will escape the local gravitational field entirely, defined by setting the kinetic energy of upward motion equal to the gravitational potential energy

Gegenschein—faint reflection of the Sun found in interplanetary dust seen opposite the Sun's position in the sky

Gas giant—see *Giant planet*

Giant planet—one of the four large outer planets of the solar system that has a fluid interior and no well-defined surface

Giant star—star that is larger than a typical star and generally has a luminosity class of III; it lies above and to the right of the main sequence of the Hertzsprung-Russell diagram

Globular cluster—large, spherical cluster of stars located in the halo of the galaxy containing 1×10^5 to 1×10^6 stars and thought to be among the oldest objects in the galaxy

Gravitational redshift—wavelength shift (toward longer wavelengths) of photons emitted by a massive body as the body's gravitational field causes energy to be lost via escaping photons

Greenhouse effect—trapping of heat near a planet's surface by atmospheric constituents that allow visible sunlight to enter the atmosphere; this also traps infrared radiation emitted by the surface as it is heated

Hertzsprung-Russell (H-R) diagram—diagram in which stars are represented according to a measure of their luminosities (commonly the absolute magnitude) on the vertical axis and a measure of their surface temperatures (commonly spectral type or color index) on the horizontal axis

H II region—volume of ionized gas in space which emits light in the form of emission lines, particularly the strong red line (H-alpha) of hydrogen

Hubble constant—constant in the Hubble law, which gives the rate of the universe's expansion; represented by H

Hubble law—mathematical expression that determines the rate of uniform expansion of the universe, given by $v = Hd$, with v being the expansion velocity of a galaxy at distance d and H being the Hubble constant

Hydrostatic equilibrium—state of balance between the inward force of gravity and the outward force of pressure that exists at all points inside any stable object, such as a star or planet

Inertia—property of matter which resists changes in motion

Inflationary universe theory—cosmological theory in which the universe is believed to have undergone, at very early times, a brief period of very rapid expansion, the precursor to the big bang expansion thought to be continuing today

Interferometry—use of interference phenomena in electromagnetic radiation to obtain precise measurements of angles in astronomical observations, utilizing two or more telescopes separated by large distances

Interstellar extinction—obscuration of light from distant stars as a result of the absorption and scattering of light by tiny solid particles of interstellar dust, thereby causing distant stars to appear dimmer

Ionization—process that removes one or more electrons from an atom, leaving a positively charged ion

Kinetic energy—energy of motion

Kirkwood's gaps—series of gaps in the asteroid belt at distances from the Sun where a body would be in orbital resonance with Jupiter; it is thought that repeated alignments of such bodies with Jupiter and the Sun have, in time, altered their orbits so that no such bodies are found at these positions today

Kuiper belt—disklike distribution of cometary bodies thought to lie in the outer solar system (between roughly 30 and 50 AU from the Sun), and a possible source of periodic comets

Local Group—group of galaxies consisting of the nearest 30 or so galaxies, including the Milky Way

Long-period comet—comet with a long orbital period that is likely to be seen only once (as opposed to a periodic comet)

Luminosity (L)—energy per second emitted by an object; measured in watts (W)

Luminosity class—one of several classes to which a star can be assigned on the basis of certain luminosity indicators in its spectrum; the classes range from I (for supergiant stars) to V (for main-sequence stars)

Lunar month—passage of time on Earth that equals the synodic period of the Moon; it is approximately 29.5 days

Magnitude—see *Stellar magnitude system*

Main sequence—the diagonal strip where most stars are located in the H-R diagram (from upper left to lower right); stars on the main sequence are thought to be in their hydrogen-burning phase

Maria—large, dark areas on the Moon's surface

Mean solar day—see *Solar day*

Meteor—streak of light observed when a small body from space enters Earth's upper atmosphere

Meteorite—solid body that reaches Earth's surface intact after falling into the atmosphere from space

Meteoroid—solid particle causing a meteor

Minor planet—see *Asteroid*

Neutrino—massless subatomic particle created in many types of nuclear reactions and thought to permeate space in enormous numbers

Neutron—uncharged particle that exists in an atomic nucleus

Neutron capture reaction—reaction in which free neutrons are captured by other nuclei to build heavier elements

Neutron star—compact stellar remnant supported by degenerate neutron gas pressure

Nova—sudden brightening of a binary system in which a white dwarf accretes mass from its companion star, causing a chain of nuclear reactions to occur; in a recurrent nova, this process might repeat itself on timescales of decades and in a dwarf nova, it might happen repeatedly in hours or days

OB association—loose grouping of young stars, including massive, hot, young stars (spectral types O and B)

Oort cloud—spherical cloud of cometary bodies thought to surround the Sun at a distance of 100,000 AU and thought to be the source of long-period comets

Open cluster—loose conglomeration of several hundred stars, commonly found in the galactic plane

Orbital resonance—situation in which the orbital periods of two bodies are simple multiples of each other and the bodies frequently align

Parsec—unit of distance defined as the distance to a star whose parallax angle is one second of an arc; one parsec is equal to 206,265 AU or 3.09×10^{13} km

Periodic comet—comet whose orbital period is short enough (a few years to a few decades) so that it is recognized as making repeated passages through the inner solar system

Period-luminosity relationship—general correlation between mass and luminosity for stars that reflects the strong dependence of luminosity on mass

Photometry—science of measuring the brightness (or magnitude) of astronomical objects

Photon—particle of light

Photosphere—visible surface layer of the Sun or a star from which continuous radiation escapes into space

Planetary nebula—tenuous, glowing shell of gas surrounding the exposed core of a star that has undergone its red-giant wind phase; the shell may be formed when a rapidly expanding stellar wind from the hot core of the star sweeps up remnant material from the slowly expanding red-giant wind

Planetesimal—small Sun-orbiting body thought to have formed in the early solar system as the precursor of planets

Potential energy—stored energy, or energy which must be released in order to perform work; it can exist in many forms, including gravitational potential energy or chemical potential energy

Precession—slow shifting of the stars' positions on the celestial sphere caused by a 26,000-year periodic wobble of Earth's rotation axis

Proton—positively charged particle that exists in an atomic nucleus

Proton-proton chain—sequence of nuclear reactions that fuse hydrogen into helium; the dominant source of energy in the Sun

Protostar—star in the process of forming; specifically, a star in the slow contraction phase prior to ignition of nuclear reactions

Pulsar—rapidly rotating neutron star that emits a periodic pulse of radio waves

Quasar—member of a class of starlike extragalactic objects characterized by an emission-line spectrum and a very large redshift; it may be the active core of a young galaxy

Quasistellar object (QSO)—see *Quasar*

Radial velocity—relative speed of an object along the line of sight between the source and the observer

Radiation pressure—form of pressure created by the absorption of photons of light

Radio galaxy—galaxy that emits a significant fraction of its radiant energy at radio wavelengths; it commonly is characterized by two lobes of radio-emitting gas

Rays—light-colored linear streaks on the Moon's surface seen radiating away from some young craters

Reflection nebula—interstellar cloud containing dust situated behind a hot star (as seen from Earth) that reflects light from the star toward Earth; it typically is characterized by a blue color because the dust grains scatter short-wavelength light most efficiently

Reflector—telescope that uses mirrors to bring light to a focus

Refractor—telescope that uses lenses to bring light to a focus

Resolution—see *Resolving power*

Resolving power—in angular measure: a measure of the sharpness of an astronomical image or map, defined as the smallest angular separation or detail that can be distinguished; in spectroscopy: a measure of the level of detail discernible in a measured spectrum, defined as the ratio of the wavelength being observed to the smallest wavelength interval that can be distinguished

Retrograde motion—apparent backward motion of an object in the sky, which is opposite to the direction of most objects in the solar system

Right ascension—equatorial coordinate that specifies the east-west position of an object in the sky; it is measured in hours, minutes, and seconds

Rilles—meandering lines on the Moon's surface that are the remains of lava rivers

Roche limit—the minimum distance that an orbiting body can exist in relation to its parent body, defined as the distance at which the tidal force acting to stretch the body apart is equal to the gravitational force acting to hold it together

RR Lyrae variable—class of periodic variable star that has a specific luminosity; used as a distance indicator

Seismic waves—vibrations caused by earthquakes

Seyfert galaxy—class of spiral galaxies characterized by an unusually bright, compact nucleus

Shepherd satellites—pair of satellites in nearby orbits whose combined gravitational effects trap ring particles between them

Sidereal day—length of time required for Earth to undergo one full rotation, as seen with respect to the fixed stars; the length of time required for a given star to make one full cycle around the sky, as seen from a fixed location on Earth

Sidereal period—length of time required for an object to undergo one orbital period or rotational period as seen by a distant observer

Solar day—length of time required for the Sun to make one full cycle around the sky, as seen from a fixed location on Earth

Solar nebula—disk of gas and dust that surrounded the Sun as it formed; the material from which the planetary system formed

Solar wind—the flow of charged particles from the Sun that permeates the solar system

Solstice—either of the two points in Earth's orbit where the Sun appears at its maximum northern or southern displacement from the celestial equator

Spectral lines—discrete lines appearing in spectra that correspond to the absorption or emission of light at specific wavelengths

Spectral type—measure of stellar surface temperature, defined by the appearance of absorption lines in the spectrum of a star which indicate the degree of ionization in the star's atmosphere

Spectrograph—instrument that records the spectra of light sources

Spectroscopic binary—binary star system characterized by spectral lines that periodically are split and recombined as a result of the Doppler effect; a spectroscopic binary may be a single-line spectroscopic binary if the light from only the brighter star is detected or it may be a double-lined spectroscopic binary if the light from both stars is detected

Spectroscopic parallax—distance to a star determined from comparison of its apparent and absolute magnitudes, with the absolute magnitude derived from the star's position on the H-R diagram, which is determined by its spectral class

Spectroscopy—study of the spectra of stars and other objects

Spectrum binary—binary star system recognized as such because its spectrum shows lines representing two distinct spectral types

Spin-orbit coupling—a simple relationship between the orbital and spin periods of a satellite or planet, normally caused by tidal forces which have slowed the rotation of the satellite or planet; the simplest case is synchronous rotation, in which the orbital and spin periods are equal

Spiral density wave—spiral wave pattern in a rotating, thin fluid disk, creating permanent phenomena, such as the rings of Saturn or the spiral arms of a spiral galaxy

Spiral galaxy—class of galaxies defined by their spiral arm structure; about half are barred spirals, having an elongated nucleus from which the spiral arms emanate

Standard candle—any object whose absolute magnitude (hence its distance) is inferred from indirect indicators, such as the assumption that it conforms to a standard relationship between absolute magnitude and some observable property

Stellar magnitude system—system for measuring stellar brightness based on the response of the human eye but quantified so that a difference of one magnitude corresponds to a ratio equal to the fifth root of 100 (2.512)

Stellar parallax—apparent annual shift in position of a nearby star as seen from Earth; this shift is a reflection of Earth's orbital motion

Stellar populations—classes to which stars in a galaxy are assigned according to various criteria, such as the abundance of heavy elements or location in the galaxy; these classes are thought to indicate the relative ages of stars

Sunspot—localized region of high magnetic field intensity on the surface of the Sun that appears darker than its surroundings because it is cooler

Supercluster—cluster of galaxy clusters

Supergiant star—star that is larger than a main-sequence star and typically is in luminosity class I; it lies at the top of the H-R diagram

Supernova—high-energy explosion that destroys most or all of a star

Synchronous rotation—see *Spin-orbit coupling*

Synchrotron radiation—electromagnetic radiation produced by electrons moving at nearly the speed of light in the presence of a strong magnetic field; it is emitted over a very broad spectral range and commonly is polarized

Synodic period—orbital or spin period of an object as seen by an observer on Earth; for the Moon or a planet, the synodic period is the interval between repetitions of the same phase or configuration

Tectonic activity—geophysical processes involving motion of crustal plates that produces such phenomena as earthquakes and volcanism

Terrestrial planet—any of the four relatively small, dense planets in the inner solar system that have rocky surfaces

Thermal radiation—continuous radiation (in the form of light) emitted from an object because its temperature is above absolute zero; its spectrum is described mathematically by the Planck function

Tidal force—difference between gravitational forces exerted on opposing sides of a body, such as a satellite or a planet that orbits close to its parent body; this differential force acts to stretch the satellite or planet along a line connecting it to its parent body

Triple-alpha reaction—reaction that fuses three helium nuclei into a carbon atom

Type I supernova—supernova that typically shows no lines of hydrogen in its spectrum and is thought to result from the nuclear detonation of a carbon white dwarf which has gained new matter in a mass-exchange binary system

Type II supernova—supernova that normally has strong lines of hydrogen in its spectrum and is thought to be the result of core collapse and rebound in a massive star that has completed its nuclear evolution and has an iron core

Van Allen belts—zones of high-density charged particles surrounding Earth, confined by Earth's magnetic field

Variable star—any star whose luminosity varies; a major class of variable stars are the pulsating variables stars, which pulsate (expand and contract) periodically

Visual binary—binary star system in which the two stars can be observed as separate objects

X-ray binary—binary system in which one star—a neutron star or black hole—has an accretion disk that emits X-rays

White dwarf—compact stellar remnant supported by degenerate electron gas pressure

Zodiac—broad strip around the celestial sphere defined by the sequence of major constellations through which the Sun appears to pass during the course of the year

Zodiacal light—faint glow along the ecliptic that is a result of sunlight being scattered by interplanetary dust

Selected References

Abell, G.D. Morrison, D., and Wolff, S.C. *Exploration of the Universe* (6th ed.). Philadelphia: W.B. Saunders Co., 1991.

Chaisson, E., and McMillan, S. *Astronomy Today:* Englewood Cliffs, N.J.: Prentice Hall, 1992.

Dixon, R.T. *Dynamic Astronomy* (6th ed.). Englewood Cliffs, N.J.: Prentice Hall, 1991.

Hartmann, W.K. *Astronomy: The Cosmic Journey* (4th ed.). Philadelphia: W.B. Saunders Co., 1989.

Kaufmann, W.J., III. *Universe* (3rd ed.). New York: W.H. Freeman, 1991.

Pasachoff, J.M. *Contemporary Astronomy* (4th ed.). Philadelphia: W.B. Saunders Co., 1989.

Seeds, M.A. *Horizons: Exploring the Universe* (4th ed.). Belmont, Calif: Wadsworth, 1993.

Snow, T.P. *The Dynamic Universe: An Introduction to Astronomy* (4th ed.). St. Paul: West Publishing Co., 1991.

Zeilik, M. *Astronomy: The Evolving Universe* (6th ed.). New York: Wiley & Sons, 1991.

Index

A
Aberration, chromatic, 40-41
Absorption line, 33, 34, 143
Accretion disk, 129
Active galactic nuclei (AGN), 158
Albedo, 85
Alpha capture reaction, 117
Alpha Centauri, 100
Andromeda, 4, 145
Angstrom unit, 30
Angular unit, 3
Anion, 36
Aphelion, 20
Appollonius, 17
Aristarchus, 17
Aristotle, 17, 20, 21
Asteroid, 54, 84-85
Asthenosphere, 59
Astrology, 2-3
Astrometry, 99-100
Astronomical unit, 5
Astronomy
   ancient history of, 15-18
   Newtonian laws and, 22-27
   Renaissance and, 18-22
Astrophysics, 1
Atom, 34
Atomic number, 94i

B
Babylonians, astronomy and, 15-16
Big bang theory, 163
Binary stars, 104-105, 106i, 125-126, 134-135
Black dwarf, 129
Black hole, 125, 134-135
BL Lacertae objects, 159
Blueshift, 37
Bohr, Niels, 29
Brahe, Tycho, 19, 20, 86
Breccias, 62
Butterfly diagram, 98

C
Callisto, 75
Cannon, Annie J., 103-104
Cassegrain focus, 42
Cassini division, 78
Cation, 36
Celestial pole, 6
Cepheid variables, 137i
Ceres, 84
Charge-coupled device, 44
Charon, 81, 82i
Chondrite, 89
Chromosphere, 91, 95, 96
Closed universe model, 163

Cloud
   dark, 144-145
   interstellar, 53
   Magellanic, 4, 154
   Oort, 55, 86-87
Cluster, stellar, 118-119, 120i
CNO cycle, 115, 116i
Color index, 112
Color-magnitude diagram, 119
Coma, 87
Comet, 86-88
Convection, 56, 95
Coordinate system
   ecliptic, 8
   equatorial, 7-8
   galactic, 8
Copernicus, Nicolaus, 19
Corona, 12, 49, 91, 96
Cosmic background radiation, 50, 164-165
Cosmology, 15
   cosmogony and, 162, 166-171
   observational, 164-166
   relativistic, 162-163
Coudé focus, 42
Crab Nebula, 130, 132

D
Dark matter, 4, 141
Declination, 8
Decoupling, 169
Deimos, 69, 72
Dicke, Robert, 164
Differentiation, 55
Diffraction, 30, 41
Dione, 78
Dipole anisotropy, 165
Distance modulus, 110, 153i
Doppler effect, 36-37, 113
Dust
   interplanetary, 90
   interstellar, 142-143

E
Earth
   atmosphere of, 58-59
   crust of, 60-61
   development of, 62-64
   interior of, 59-60
   motions of, 6-10
   rocks of, 60
   solar system and, 3
Eclipse
   lunar, 12, 13i
   solar, 11-12, 13i
Ecliptic, 8
Einstein, Albert, 29, 162-163

Electromagnetic wave, 29-30, 31i
Electron, 34
Emission line, 33, 34
Emission nebula, 144
Enceladus, 77
Energy, stellar, 24-25, 117-118
Epicycle, 17
Equator, celestial, 8
Equilibrium, hydrostatic, 55, 115
Equinox, 9
Eratosthenes, 17
Escape speed, 24-25
Europa, 75
Excitation, 35-36
Extinction, interstellar, 142

F
False vacuum, 169
Flare, solar, 97
Flat universe model, 163
Force
   electromagnetic, 167
   gravitational, 27
   nuclear, 167
Frequency, 30

G
Galaxies
   clusters of, 154-155
   elliptical, 150-151
   groups of, 154
   properties of, 152-154
   radio, 158, 159
   Seyfert, 159
   spiral, 150-151
Galileo (astronomer), 21-22, 39, 73, 76
Galileo (space probe), 51, 74, 84
Galle, Johann, 79
Gamma ray, 30, 50
Gamow, George, 164
Ganymede, 75
Gas, degenerate
   electron, 122
   neutron, 125
Gas giants, 52, 73-81
Gaspra, 84
Gegenschein, 90
Giant star, 108
Gravitation, universal, 24-25
Gravity, 3-4, 167
Great Red Spot, 74
Greeks, astronomy and, 16-18
Greenhouse effect, 68

i refers to an illustration

182 Index

H
Halley, Edmund, 86
Halo, 87, 118
Helium flash, 123
Herschel, William, 79
Hertz, 30
Hertzsprung, Ejnar, 108
Hertzsprung-Russell diagram, 108, 109i, 110, 114, 120i
Hipparchus, 17-18, 101
H II regions, 144
Hubble, Edwin, 150-152
Hubble law, 156, 157i
Hubble Space Telescope, 46
Huygens, Christiaan, 29, 76
Hyperion, 77

I
Ida, 84
Inertia, 21, 23
Inflationary universe theory, 169-170
Infrared radiation, 30, 48-49
Interference, 30
Interferometry, 46, 100
Interplanetary bodies, 54-55, 84-90
Inverse beta decay, 168
Io, 75
Ionization, 35-36, 92
Isotope, 94i

J
Jansky, Karl, 45
Jupiter
  asteroids and, 84-85
  atmosphere of, 74
  interior of, 74
  magnetic field of, 74-75
  particle belts and, 74
  satellites of, 75-76

K
Kelvin scale, 32
Kepler, Johannes, 19-20, 21i, 110-111
Kirchhoff's rules, 33
Kirkwood's gaps, 85
Kuiper belt, 55, 87

L
LeMaître, Georges, 163
Light, radiation and, 29-37
Light-year, 5
Lithosphere, 59
Local Group, 154, 165
Luminosity, 32-33, 105, 107-108

M
Magellan (space probe), 68
Magellanic Clouds, 4, 154
Magnetic field, 97, 113
Magnitude, stellar, 18, 101-102, 107-108
Main sequence, 108, 119
Maria, 61
Mariner (space probe), 65, 67, 70
Mars, 69
  atmosphere of, 70
  moons of, 72
  orbit of, 20
  search for life on, 71-72
  surface of, 70-71
  water on, 71
Mass, 23, 25, 26i
Maunder Minimum, 98
Mean solar day, 7
Mercury
  interior of, 67
  orbit of, 65, 66i
  surface of, 66-67
Mesosphere, 59
Meteor, 88
Meteorite, 89
Meteoroid, 88
Milky Way, 3-4, 154
  formation and evolution of, 145-149
  interstellar components of, 141-145
  structure of, 136-138, 139i, 140-141
Mimas, 78
Miranda, 80
Molecule, 36
Momentum, angular, 25, 62-63
Moon
  configurations of, 10-11
  development of, 62-64
  eclipse of, 12, 13i
  interior of, 62
  phases of, 10-11, 12i
  sidereal period of, 10, 11i
  surface of, 61-62
  synodic periods of, 10, 11i
Motion, laws of, 20, 21i, 23-24

N
Nanometer, 30
National Aeronautics and Space Administration, 2
National Science Foundation, 2
Nebula
  emission, 144
  planetary, 123
  reflection, 144
  solar, 53

Neptune
  atmosphere of, 79-80
  interior of, 80
  moons of, 80
  rings of, 80-81
Nereid, 80
Neutrino, 93, 168
Neutron, 34, 168
Neutron capture reaction, 118
Neutron star, 125, 131-132, 133i, 134
Newton, Sir Isaac, 1, 22-26, 29, 39
Nova, 129

O
OB association, 119
Observatories, 44, 47-50
Occam's razor, 2
Oort cloud, 55, 86-87
Open universe model, 163
Orbital resonance, 75, 85

P
Parabola, 41
Parallax
  spectroscopic, 110
  stellar, 17, 100, 101i
Parsec, 5, 100
Particles, radiation and, 29-31
Penzias, Arno, 165
Perihelion, 20
Permafrost, 71
Phobos, 69, 72
Photoionization, 35
Photometry, 42, 99, 100-102
Photomultiplier, 42
Photon, 31, 168
Photosphere, 95
Pioneer (space probe), 67, 68, 74
Pixel, 43
Planck, Max, 29
Planck function, 32
Planetesimal, 54
Planets
  atmosphere of, 56
  configurations of, 13
  formation of, 54-55
  giant, 52, 73-81
  minor, 54, 84
  motions of, 14, 20, 21i, 52
  nebula of, 123
  structure of, 55-56
  terrestrial, 52, 56
Plato, 16-17
Pluto, 54
  atmosphere of, 82-83
  Charon and, 81, 82i
  interior of, 82-83
  rotation of, 81, 82i

i refers to an illustration

Index

Polaris, 10
Polarization, 31
Population I and II stars, 145-146
Positron, 168
Precession, 9-10, 18
Proton, 34, 168
Proton-proton chain, 93, 94i, 115
Protostar, 53, 121
Ptolemy, 18
Pulsar, 132, 133i
Pythagoras, 16

Q

Quadrature, 10, 12i
Quanta, 31
Quarks, 167
Quasars, 158, 159-161

R

Radiation
  blackbody, 32
  continuous, 31-33
  cosmic background, 164-165
  electromagnetic, 29-31
  synchrotron, 130
  thermal, 32-33
Radiation pressure, 87, 116
Radio galaxies, 158, 159
Radio waves, 30, 46-47
Rays, lunar craters and, 62
Reber, Grote, 45
Recombination, 169
Redshift, 37, 128, 158, 160
Reflection nebula, 144
Refraction, 30
Regolith, 62
Relativity, 162-163
Renaissance, astronomy and, 18-22
Resonance, orbital, 75, 85
Retrograde motion, 14
Rhea, 77
Right ascension, 8
Rilles, 62
Roche limit, 78
Rocks, types of, 60
Rotation, synchronous, 27, 61
RR Lyrae variables, 138
Russell, Henry Norris, 108
Russell-Vogt theorem, 114

S

Sagittarius A, 141
Satellite
  co-orbital, 77-78
  shepherd, 78
Saturn
  atmosphere of, 76-77

Saturn *(continued)*
  interior of, 77
  moons of, 77-78
  rings of, 78-79
Schwarzschild radius, 135
Seasons
  on Earth, 8-9
  on Mars, 70
  on Uranus, 79
Seismic wave, 59
Seyfert galaxies, 159
Sidereal day, 7i
Sidereal period, 10, 11i
Sirius, 102, 127
Solar day, 7i
Solar system
  formation of, 52-55
  overview of, 51-52
Solar wind, 87, 96
Solstice, 9
Spectral line, 33-34, 35i, 36
Spectrograph, 42
Spectroheliogram, 96
Spectroscopy, 99, 102-104
Spectrum, electromagnetic, 29-30
Spin-orbit coupling, 66i
Spiral density wave, 78, 146-147
Standard candles, 152, 153i
Star
  binary, 104-105, 106i, 125-126, 134-135
  brightness of, 100-102
  clusters of, 118-119, 120i
  composition of, 112-113
  energy of, 24-25, 117-118
  evolution of, 121, 122i, 123-125
  formation of, 120-121
  giant, 108
  Hertzsprung-Russell diagram of, 108, 109i, 110, 120i
  luminosity of, 105, 107-108
  magnetic field of, 113
  magnitude of, 18, 101-102, 107-108
  mass of, 110-111, 114-115
  neutron, 125, 131-132, 133i, 134
  parallax motion of, 17, 100, 101i
  population I and II, 145-146
  position of, 99-100
  radius of, 111-112
  rotation of, 113
  spectra of, 102-104
  structure of, 114-118
  supergiant, 108
  temperature of, 104, 112

Star *(continued)*
  variable, 104, 137i, 138
Stefan-Boltzmann law, 32-33
Stratosphere, 59
Sun
  atmosphere of, 95-96
  eclipse of, 11-12, 13i
  energy and, 93-95
  formation of, 53
  internal structure of, 92-93
  magnetic field of, 97
  properties of, 91-92
  seasons and, 8
Sunspot, 97-98
Supercluster, 4, 155
Supergiant star, 108
Supernova, 125, 130-131
Synodic period, 10, 11i

T

Tectonic activity, 60
Telescopes
  advantages of, 39-40
  gamma-ray, 50
  infrared, 48-49
  observatory sites and, 44, 47-50
  radio, 46-47
  reflecting, 41-42, 43i
  refracting, 40-41
  resolving power of, 40, 41
  ultraviolet, 47-48
  X-ray, 49
Terrestrial planets, 52, 56
Tethys, 78
Thales, 16
Theories, scientific, 2
Thermosphere, 59
Tides, 27, 61
Timekeeping, astronomy and, 16
Titan, 77
Tombaugh, Clyde, 81
Triple-alpha reaction, 117, 118i
Triton, 80
Trojan point, 78
Troposphere, 59
Tuning fork diagram, 151, 152i

U

Ultraviolet radiation, 30, 47-48
Umbra, 13i
Universal gravitation, law of, 24-25
Universe
  age of, 156-157
  closed, 163
  expansion of, 156-158, 165-169
  future of, 170-171

i refers to an illustration

Universe *(continued)*
   Hubble law and, 156, 157i
   inflationary, 169-170
   open, 163
   redshifts and, 158
   size of, 4-5
   structure of, 3-4
Uranus
   atmosphere of, 79-80
   interior of, 80
   moons of, 80
   rings of, 80

V
Vacuum, false, 169
Van Allen belt, 56, 60
Variable stars, 104, 137i, 138
Velocity, radial, 37
Venus
   atmosphere of, 68
   evolution of, 69
   interior of, 68-69
   surface of, 68-69
Viking (space probes), 70, 71
Volcanoes, 60-61
Voyager (space probes), 74-76, 80-81

W
Wave
   electromagnetic, 29-30, 31i
   radio, 30, 46-47
   seismic, 59
   spiral density, 78, 146-147
White dwarf, 108, 123, 127-130
Wien's law, 32
Wilson, Robert, 165
Wind
   solar, 87, 96
   stellar, 116

X-Z
X-ray binaries, 134-135
X-rays, 30, 49
Zeeman effect, 129
Zodiac, 8
Zodiacal light, 90

i refers to an illustration